The Nile Basin

The Nile Basin

National Determinants

of Collective Action

John Waterbury

Yale University Press

New Haven and London

Set in Adobe Garamond type by The Composing Room of Michigan, Inc.
Printed in the United States of America by Sheridan Books, Ann Arbor, Michigan.

Library of Congress Cataloging-in-Publication Data

Waterbury, John.
    The Nile Basin : national determinants of collective action / John Waterbury.
        p.   cm.
    Includes bibliographical references and index.
    ISBN 0-300-08853-1 (cloth : alk. paper)
        1. Water-supply—Nile River Watershed—Management—International
cooperation.   2. Water resources development—Nile River Watershed—
International cooperation.   3. Water resources development—Government
policy—Nile River Watershed.   I. Title.

    TD319.N6 W38   2001
    333.91'15'0962—dc21

                                                                    2001045488

A catalogue record for this book is available from the British Library.

The paper in this book meets the guidelines for permanence and durability of
the Committee on Production Guidelines for Book Longevity of the Council
on Library Resources.

10   9   8   7   6   5   4   3   2   1

Contents

O. H.

Preface

Generations of political scientists have adopted Harold Lasswell's famous dictum that politics is the study of who gets what, when, and how. Politics is indeed about power and the distribution of goods—public, private, and psychic. One could posit just as parsimoniously that politics is the study of the search for collective action. Thomas Hobbes performed thought experiments on the state of humankind before collective action and imagined the war of all against all. In that imagined world there were no alliances, no friends, no laws, no tacit agreements. All were enemies, and the only "incentives" at hand were sanctions, presumably of the deadliest kind. We do not know when humankind first consciously sought to form communities, to build positive incentive systems to bring about voluntary cooperation, and to contain the brutish violence and destruction of the Hobbesian construct. Yet that quest encompasses all the political history of human communities. The logic of collective action, to use the phrase brought to prominence by Mancur Olson, has not qualitatively changed throughout thousands of years of our recorded experience as social animals.

Because the quest is ongoing and the quarry elusive, the fundamental questions that have challenged, and often defeated, generations of political philosophers, political economists, and latterly, game theorists, are still very much with us. The challenge is so generic that we now summarize it in the literature as a CAP—collective action problem. There are many paths to approaching it. Any collectivity, from a student club to a trade union or parties to an international convention, faces the same internal dynamics if cooperation and collective action are to be achieved and sustained.

I have chosen the Nile basin, with its ten riparian nations, as the focus for my own attempt to understand these dynamics in an international setting. I use the word *chose* somewhat loosely. The Nile chose me during a long period of residence in Egypt in the 1970s (see Introduction). The challenge is establishing a cooperative framework, followed by collective action, in the joint distribution and management of a scare and irreplaceable resource—the waters of the Nile and its tributaries. "Common property" resources, such as the Nile, are as old as humankind. Water and pasturage have throughout history been at the heart of efforts to coordinate use. These efforts frequently fail. Cooperation is hard to achieve and even more difficult to sustain, and when efforts collapse, conflict and resource destruction often are the consequences. We slip backward toward the Hobbesian construct.

The twenty-first century is likely to be one in which questions of coordinated and managed use of scarce common-property and common-pool resources (see Chapter 1) are dominant in the international or global arena. The pressures generated by billions of people on the atmosphere, the oceans, arable land, and fresh water supplies are already the subject of international concern but, as yet, little international action. Why that is so, I think, can be accurately and usefully explained through a detailed examination of the quest for cooperation in the Nile Basin. The case itself is intrinsically important, involving the welfare of hundreds of millions of inhabitants in northeast and equatorial Africa. It is one, very big, piece of a global challenge.

ACKNOWLEDGMENTS

I have thirty years of debts for which to account insofar as my understanding of the Nile Basin and collective action problems are concerned. To some, like the late Mancur Olson, the debt is purely intellectual and based on the corpus of his written work (although we did have a wonderful dinner together in Mexico City several years ago). I owe that same debt of close reading and inspiration in varying ways to Elinor Ostrom, Marc Lichbach, Oran Young, Peter Haas, Douglass North, and to specialists in international law such as Stephen Mc-Caffrey and Albert Utton.

Three anonymous reviewers of this manuscript for Yale University Press provided valuable advice which, I hope, they will find reflected in the final book.

To others the debts are both intellectual and personal and include Asit K. Biswas, John Kolars, Aaron Wolf, Dale Whittington, Miriam Lowi, Robert Bates, James Scott, Tony Allan, Peter Rogers, Charles Okidi Odidi, Steven Brichieri-Colombi, Elizabeth Wackett, and Senai Alemu.

Much of the academic year 1995–96 was spent in Ethiopia and Uganda. I hereby acknowledge my great debt to the Fulbright Exchange Program, which supported my research throughout that academic year. It was a sequel to a much earlier Fulbright grant for language study in Egypt in 1961. I owe more than I can say to this great initiative of the late Senator William J. Fulbright.

Many people in Ethiopia and Uganda gave graciously of their time, expertise, and documentation. In Ethiopia, Prime Minister Meles Zenawi discussed Nile issues with me on two occasions. Others in Ethiopia who were of great help to me were Imeru Tamrat, Senai Alemu, Kifle Wadajo, Solomon Kasahun, Tekeda Alemu, Gerard Chetboun, Daniel Gamacho, Adam Ereli, Alem-Tsehai Asfaw, and Suzanne Jacquette. In Uganda, Geoffrey Onegi Obel, Patrick Kahangire, Enoch Dribidu, and John Ntambirewki are all gratefully acknowledged.

Pagan Amum, special advisor to John Gareng, leader of the SPLA/SPLM, put me in touch with several southern Sudanese resistance leaders and helped me attend a SPLM conference in "New Cush."

Experts in the World Bank also gave generously of their time. I am especially grateful to David Grey, Tony Sparkes, Fayyez Omar, Nadendra Sharma, and Aktar Elahi.

I did not carry out fresh field research in either Egypt or the Sudan. I have kept abreast of developments in Nile basin issues in both countries from written sources and occasional visits. My extensive acknowledgments in *The Hydropolitics of the Nile Valley* to colleagues in both countries are still valid. I would add only that anyone concerned with Nile basin issues must be reassured that Mahmoud Abou Zeid is now (summer 2001) Egypt's Minister of Public Works and Water Resources.

I am grateful to Stephanie Allen for her meticulous help in preparing the manuscript for submission to Yale University Press and to Noreen O'Connor for her fine editing.

Abbreviations and Terms

Abbay	Ethiopian name for the Blue Nile
asl	above sea level
bcm	billion cubic meters
belg	the "short" rains of March–April in the Ethiopian highlands
cumsec	cubic meters per second
dergue	Ethiopian junta, 1974–91
ECA	Economic Commission for Africa
EELPA	Ethiopian Electric Light and Power Authority
EPDRF	Ethiopian Peoples' Democratic Revolutionary Front
EVDSA	Ethiopian Valleys Development and Study Authority
FAO	Food and Agriculture Organization
feddan	1.03 acres (Egypt, Sudan)
hectare	2.47 acres
Hydromet	Hydrometeorological Survey of the Catchments of Lakes Victoria, Kyoga, and Mobutu
ICCON	International Consortium for Cooperation on the Nile

IGADD	Intergovernmental Authority on Drought and Development
ILC	International Law Commission
KBO	Kagera Basin Organization
kWh	kilowatt hour
m³	1,000 liters; 1 metric ton; 35.3 cubic feet
mcm	million cubic meters
mehr	the "long" rains of July–September in the Ethiopian highlands
mt	metric ton
MW	megawatt
negus	the Amharic title of Emperor
OAU	Organization of African Unity
PJTC	Permanent Joint Technical Commission on the Nile
RBO	river basin organization
SPLA	Southern People's Liberation Army
SPLM	Southern People's Liberation Movement
Teccaze	Ethiopian name for the Atbara River
Tecconile	Technical Cooperation Committee for Promotion of Development and Environmental Protection of the Nile Basin
UAR	United Arab Republic (Egypt)
UEB	Uganda Electricity Board
UNDP	United Nations Development Program
UNEP	United Nations Environmental Program
USAID	United States Agency for International Development
WHO	World Health Organization
WMO	World Meteorological Organization

Introduction

Ten juridically sovereign nations share the Nile basin and by virtue of that fact can lay claim to *some* share of its waters. Why they would do so and how they would go about it are questions shaped by relative power among the states, access to patrons external to the basin, and access to alternative sources of water. No state in the basin will forswear its claims simply because it has little interest in the water itself. We may place this reality alongside one equally as important: some states in the basin are highly dependent in fundamental ways upon the water, whereas others want to become so dependent. As it becomes increasingly evident that the supply of Nile water is not likely to meet all future demands—given the technologies with which and the patterns by which it is currently used—not only must those states most concerned with supply reach some sort of equilibrium of supply and demand among themselves, but they must also accommodate in some fashion those least concerned.

The equilibrium may be struck on the basis of relative power and resources, and on the basis of national interests shaped, to date, by state policy-making elites as yet little influenced by domestic economic interests. The language in which the equilibrium is negotiated is that of "appreciable harm" and "equitable use." These two terms dominate the international law debate over the management of international watercourses (see Chapter 1), but they are not mere legalisms. They in fact capture in summary form the respective interests of downstream states, which generally have the most ancient or senior claims to "rights" in the use of the water, and of the upstream states, which wish to assert such claims. In the Nile basin the downstream states, principally Egypt and the Sudan, defend what they have in the language of appreciable harm. The upstream states, represented most forcefully by Ethiopia, assert their claims in the language of equitable use.

HAS THE UPPER BASIN'S MOMENT COME?

This book began in 1974. At that time, as Middle East correspondent of the American Universities Field Staff, resident in Egypt, I wrote a number of articles examining Egypt's agricultural strategy, prospects for food security, and agricultural pricing policies, including charges for irrigation water. The reservoir of the Aswan High Dam had just reached its optimal surface level after slowly filling for about a decade (1964–74), and the full range of its benefits and side effects was under active assessment, both technically and journalistically.

Egypt's agricultural sector is in many ways a paragon of simplicity; there are no frosts or freezes, no mountains or steppes, and no rain. There is no rangeland, forest, or scrub. There are no remote agricultural areas. All of the water that sustains agriculture comes from the Nile River or from the aquifers that are directly recharged from the Nile. The complexity in Egypt's agricultural system lay and lies in its control structures, water release patterns, the "landscaping" of the watercourse itself, and the complex grid of main canals, feeder canals, field canals, and drainage channels, tens of thousands of kilometers in aggregate length, that in many ways render agricultural production about as predictable as anywhere outside of hothouses. The Aswan High Dam removed the threat of floods and, so far, has guaranteed nearly all the water the agricultural sector requires, even in years of low river flow.

That level of predictability was already evident in the early 1970s, but, like a few others at the time, I asked myself, what if the flow of the Nile became an element of major uncertainty for Egypt? What if its share, rooted in practices dating back millennia, was not secure? What if the several states (riparians or those along the banks—*rives*—from whence comes our word *rival*) were to assert claims to some substantial share of the water in the system and put those claims into effect? In the early 1970s it appeared that that kind of scenario was most likely to come to pass in the Sudan.

In the early 1970s, at a time when the United States was concluding its first major sale of wheat and other basic grains to the Soviet Union, Middle Eastern nations became concerned that the large exporters of basic grains (at the time, the United States, Canada, Argentina, and Australia, and to a lesser extent the European Community)[1] would jerk Middle Eastern importers around by colluding on prices. This was a time, it should be recalled, when several Arab countries were closely aligned with the Soviet Union and might therefore expect some retribution from the mainly western, pro-American grain exporters. Even oil exporters more closely allied with the United States, such as Saudi Arabia and Kuwait, could not count on benevolent behavior on the part of the West as they (mainly prerevolutionary Iran and Saudi Arabia) had engineered the enormous increases in the price of petroleum of 1973 and 1974 that wreaked havoc in western economies.

There developed a regional concern with food security and increased agricultural production so that, collectively, the Middle East would be less reliant on imported grains and agricultural products. The 1974 United Nations Conference on Food, chaired by Egypt's long-serving Minister of Agriculture and Agrarian Reform, Sayyid Mar'ei, was shot through with this subtheme. It was, in the end, a quixotic quest, as what we now know as the Dutch Disease ran through the Middle East, shifting prices in favor of the nontraded sectors and away from the productive sectors, above all agriculture.[2] Be that as it may, a consensus developed in the Arab world that the Sudan was the last remaining frontier of agricultural development in the region. This country is Africa's largest, and the size of the United States east of the Mississippi. It seemed to hold out the promise of vast expanses of unexploited land, suitable for both rainfed and irrigated agriculture. It could produce the exportable surpluses that the rest of the Arab world would rely upon to meet their production gaps. The Sudan is too hot to produce wheat efficiently, and what wheat it does produce is on irrigated perimeters, which is a poor use of scarce water. Instead, the Sudan was targeted to produce sugar, pulses, sorghum, rice, cattle, and sheep for export. The Kuwaitis and the Saudis began to invest heavily in Sudanese agriculture. The Arab League of States created an Arab Authority for Agricultural Investment and Development and headquartered it in Khartoum, the capital of the Sudan. With international investment putting wind in their sails, the Sudanese talked of bringing a million or so hectares (1 hectare = 2.47 acres) under irrigated cultivation within ten to fifteen years, and more was to follow (see Waterbury, 1976).

Egypt simultaneously was driving ahead with its own ambitious plans for desert reclamation and irrigated production on one million or more feddans (1 feddan = 1.03 acres). The lower basin of the Nile, comprised of Egypt and the Sudan, looked likely to generate demands for water that could not be accommodated by existing average river discharge nor with current farming and irrigation practices. Something had to give, or so it seemed.

My 1979 book, *The Hydropolitics of the Nile Valley,* examined the different pieces of the water use puzzle that was producing the resource crunch. I prudently noted, however, that if either Egypt or the Sudan fell into economic crisis, their ability to implement costly irrigation and reclamation schemes would be sharply reduced. By 1977, the Sudan had begun to enter such a crisis and by the early 1980s it had reached dramatic proportions, leading to defaults on its obligations to the International Monetary Fund and the World Bank. Its existing irrigation infrastructure began to suffer badly, most of the 1970s projects were put on hold, and its demand for and use of Nile water probably declined. The supply crisis of the lower Nile basin was thus postponed.

Nonetheless, I was more than curious to know what the Upper Nile riparians were planning with respect to their water resources in the Nile basin. As the source of 80–85 percent of the Nile's annual discharge, Ethiopia was and is by far the most important actor, but Uganda (through which flows nearly all the water that is ultimately carried by the White Nile) is also of great geostrategic significance. The other upper-basin riparians—Kenya, Tanzania, Burundi, Rwanda, Zaire (since 1997, the Congo)—had potential legal rights to Nile water (see Chapter 1)—but little practical interest in Nile issues. Once having attained its independence from Ethiopia in the early 1990s, Eritrea became an official Nile riparian with only a modest stake in basin-wide cooperation.

Let us review briefly the "stakes" of the ten riparians in the Nile basin and in cooperation (see Alemu, 1995b, for a detailed analysis).

- Egypt is by far the nation most dependent upon the Nile and with by far the highest stake in whatever regime comes to prevail. Egypt is capable of imposing a preferred solution.
- Sudan is somewhat less dependent upon the Nile as it has extensive rainfed areas within its borders. However, the Sudan is heavily dependent upon agriculture, and the most reliable production areas are the irrigated surfaces in the Nile basin.

- Ethiopia at present is not dependent upon the Nile in any important respect, but it would like to be. It sees its western Nile watershed as a crucial resource for irrigated agriculture, the exploitation of which has been unfairly denied it by the prevailing regime in the basin. That regime has been established by Egypt and the Sudan.
- Eritrea's stake in the Nile is confined to management of two seasonal streams that flow from its territory into the Sudan. There have been periodic understandings between these two countries on the use of these flows. One of them now forms part of the border between Eritrea and Ethiopia, but the watercourse itself (the Mereb-Gash basin) has not been the cause or object of the fighting between Ethiopia and Eritrea.
- Kenya enjoys a portion of the shore of Lake Victoria, in the Nile basin, and a number of tributaries to Lake Victoria arise on Kenyan territory. Kenya has seen itself more as a broker in the Nile basin and has never exhibited much interest in any binding accords on water use. It does participate in the Lake Victoria Basin group of states, which is concerned mainly with issues of pollution and the spread of the water hyacinth (see Chapter 7).
- Tanzania is a member of the Lake Victoria Basin group as well as of the Kagera Basin Organization. The Kagera River is the furthest tributary of the Nile system. Tanzania does not have any major stake in issues of water supply. However, were it to pursue an old colonial scheme involving a canal from Lake Victoria to the interior of the country, Tanzania would certainly capture the attention of all the riparians in the White Nile system.
- Burundi is a member of the Kagera Basin Organization, and that river rises on its territory. Burundi has high and fairly evenly spread rainfall so that its interest in the Kagera is confined mainly to hydropower generation.
- Rwanda also is a member of the Kagera Basin Organization and, like Burundi, enjoys high and regular rainfall. It too is interested mainly in hydropower generation which, in itself, does not affect supply.
- Congo (formerly Zaire) has seldom shown any interest in the Nile. However, when Mobuto Sese Seiko was in power, he entertained the Egyptian proposal to build a vast power grid leading from Zaire's Inga hydropower station across to the Nile basin and then on to Europe. Part of the Congo's border with Uganda runs down the middle of Lake Albert, which is part of the Nile basin. The Congo has some interest in shipping and fishing rights on and in Lake Albert.
- Uganda, with Egypt, the Sudan, and Ethiopia, rounds out the quartet of major stakeholders in the Nile basin. Except in its semi-arid northeast, Uganda

does not need surface irrigation or additional surface water. However, it straddles the exit of the Victoria Nile from Lake Victoria. At Jinja, near the capital, Kampala, it operates, in direct cooperation with Egypt, the Owen Falls Dam that regulates flow from Lake Victoria and generates power. Uganda is bound by treaty to Egypt in the operation of the dam, it is a member of the Kagera Basin Organization, and it is a member of the Lake Victoria Basin group.

Whatever the degree of interest or concern, several of these countries have been in almost perpetual political and economic turmoil. A Marxist military dictatorship, lead by Col. Haile Meriam Menguistu and a junta called the *dergue,* had led Ethiopia since 1974, when it overthrew the pro-western Emperor, Haile Selassie. It had been attacked by Somalia, had begun to fight insurgencies in Eritrea and Tigray provinces, and had been shunned by the West and most international sources of credit. It was in no position to assert credible claims to Nile water.

Uganda was under the sway of the erratic and brutal Idi Amin, who drove out Uganda's Indian community, the backbone of commerce and industry. The economy was in shambles as a result. The advent, in 1981, of Milton Obote, with Tanzanian military backing did nothing to restore political or economic stability in Uganda. Rwanda and Burundi staggered along the brink of ethnic conflict, to which Rwanda succumbed in horrific fashion in 1994–95, but, blessed with abundant rain, they were interested in Nile basin water only as a source of hydropower. Zaire (in 1997, once again, the Congo) lived under the plundering hand of Mobutu Sese Seiko, and with the Zaire River, the discharge of which is five times the Nile's, had little interest in its potential Nile water resources. Only Kenya and Tanzania had shown a modicum of political stability, and only Kenya had achieved sporadic and short-lived bursts of economic growth. In sum, the upper basin riparians posed no threat whatsoever to Egypt's and the Sudan's exclusive exploitation of the Nile. That fact became abundantly clear during my own visits to the upper basin states (Ethiopia, Kenya, Tanzania, Zaire, and Uganda) during 1981–82 (see Waterbury, 1982).

A decade later the situation had begun to change dramatically in the most important riparian states: Ethiopia and Uganda. The Menguistu regime lost out in its struggle with a loosely allied group of Tigrayan, Eritrean, and Oromo revolutionaries. The collapse of the Soviet Union in 1989 meant that these leftist forces knew they had to reach an accommodation with the West if they were to rebuild Ethiopia. Moreover, Meles Zenawi, the Tigrayan leader who established the Democratic Republic of Ethiopia, almost immediately put in train a process by which the troubled province of Eritrea became a sovereign nation through referendum, and he had drafted a constitution to construct an ethnically defined federation for the rest of Ethiopia.

Similarly, in Uganda, Yoweri Museveni led a successful insurrection against Milton Obote and established a new regime that embraced an array of economic reforms recommended by the international donor community. In an equally significant move, he welcomed back and restored property to the Asian families whom Idi Amin had driven out fifteen years earlier. By the early 1990s, Uganda's GDP growth rates were among the highest in the world. Uganda inched toward a more democratic system.

In the academic year 1995–96, with a Fulbright research fellowship in hand, I went to Ethiopia and Uganda, with side trips to Kenya, Eritrea, and the southern Sudan, to take stock of riparian water policies in a new era of economic restructuring and political stability. It seemed to me that these two countries could plausibly seek international financing for hydraulic projects in hydropower and irrigation that might have significant consequences downstream. It was about this time that the leaders and governments of Eritrea, Ethiopia, Uganda, and, after the end of the near-genocide of the Tutsi in Rwanda, the new government of Paul Kigame in Rwanda, were hailed as a new force of young leaders with new agendas of market-led growth and, one day, democratic institutions. They sounded the death knell of the old guard, corrupt and rapacious leaders in the style of Mobutu, Obote, and implicitly Moi of Kenya.

The United States, emerging from a bloodying in Somalia, from costly and desperate efforts to alleviate famine throughout the Horn of Africa, and from witnessing the slaughter in Rwanda, fastened on this leadership as the foundation for stability and growth in the Horn and African Great Lakes regions. In addition, it was leadership that wanted to contain the militantly Islamic regime, led by General Omar Bashir and his mentor Hassan Turabi, that had taken power in the Sudan in 1989. There was a possibility that U.S. policy objectives in the Horn and Lakes regions might even have led it to back Upper Basin demands for use of Nile water, in the name of famine prevention and economic stability. Such a shift in U.S. policy would have run counter to the interests of Egypt, the other, long-established ally of the United States in the Nile watercourse.

That premise or hypothesis has, at the time of writing, proven sadly and badly misconceived. The new leaders either provoked or allowed themselves to be drawn into regional conflicts that looked to be unwinnable and, in any event, were ruinously costly. Uganda, Ethiopia, and Eritrea, beginning in 1996, formed an unannounced alliance to harass the regime in the Sudan by support-ing various elements of the Sudanese opposition, especially the Southern Peo-ple's Liberation Army (SPLA) of John Garang. Rwanda and Uganda lent sup-port to Laurent Kabila, who ousted Mobuto Sese Seiko from Zaire. Both efforts had the discreet backing of the United States, and both appeared to have achieved some initial success. The advent of the millennium, however, con-firmed that the new leaders were no better endowed with vision and statesman-ship than their predecessors. Rwanda quickly tired of Kabila, who failed to po-lice the Congo's border with Rwanda (Kabila had reintroduced the use of Congo as the country's official name). The Sudan succeeded in destabilizing Uganda through support of the Lord's Resistance Army in the north as much as Uganda had succeeded in destabilizing the Sudan through support of the SPLA. A famine on the scale of those of the mid-1980s and early 1990s engulfed the peoples of the southern Sudan. In the early summer of 1998, the erstwhile al-lies, Eritrea and Ethiopia, engaged in a nasty war over disputed border areas that raged on for two years. In the spring and summer of 2000, Rwandan and Ugandan troops engaged in bloody clashes in Kisangani and other parts of the Eastern Congo. The quest for stability and growth in the Horn and Great Lakes region of Africa seemed as distant at the end of the 1990s as it had at the begin-ning. Egyptian national interests in Nile water were once again protected by the political ineptitude of the upstream riparians.

WATER AND COOPERATION

The analytic issues I want to address in this book are not, however, dependent upon the regional politics described above. In the next two chapters I will lay out in some detail the central issue, which is understanding under what cir-cumstances ten sovereign riparian states would ever *voluntarily* agree to manage their shared Nile water resources for the greater good of all the inhabitants of the watercourse. In thinking about the specifics of the Nile basin, we should gain some leverage over the fundamental issue of how voluntary compliance to any set of cooperative arrangements is ever brought about.

Since my initial foray into Nile water challenges in the 1970s, there has been a spate of literature on transnational water and potential conflict over its dispo-sition. Some of this was triggered by the Israeli-Arab dispute and the Madrid negotiating process, begun after 1991, which gave great prominence to water re-sources "shared" by Jordan, Israel, and Palestine. Secretary-General of the UN

Boutros Boutros-Ghali predicted that water would be a source of international conflict, and pundits like Joyce Starr escalated the rhetoric in writing about water wars. In more rigorous scholarly treatment, Thomas Homer-Dixon (1994) warned in a general neo-Malthusian way that we should expect growing conflict over resources of all kinds as a combination of population and economic growth puts intense pressure on our stock of natural assets. In the terms I will use in this book, the proposition flowing from these analyses was that if sovereign nations could not find frameworks for voluntary adherence to cooperative management of a scarce resource, then war, or its threat, is the likely alternative.

In 1898, a small French expeditionary force, led by Captain Marchand, reached Fashoda on the northern edge of the Sudd swamps in southern Sudan. To Lord Salisbury, the British Prime Minister, and to General Kitchener, the Governor of Egypt, the French presence at the "headwaters" of the White Nile was intolerable. A superior British force obliged Captain Marchand to withdraw before any combat took place (Collins, 1990: 49–63). Like all colonial enterprises, the British were thickly populated with engineers, yet the simple engineering facts of the situation were apparently lost on British officialdom as a whole. Even had the French flag flown for sometime at Fashoda, no amount of French engineering genius could have impeded the flow of the Nile. The British reaction was unrelated to any conceivable French action.

As we shall see in Chapter 3, Great Britain was equally apprehensive about Egypt's supply of water coming from Lake Tana and the Blue Nile watershed in Ethiopia. Until World War II, when maintaining the empire became politically and economically unfeasible, Great Britain continually sought indirect control over the upper Blue Nile watershed. Yet from 1882 on, when Great Britain took formal control of Egypt, either the flag of Ethiopia's Emperor or that of the Italians flew over the highlands. In all those years and up to the present time, only two minor hydraulic works were constructed in the Ethiopian Blue Nile watershed: a small hydroelectric power station at the outlet of the Blue Nile from Lake Tana and a somewhat larger hydropower station and reservoir on the Finchaa tributary. What we may call "water security" issues provoke misconceived and disproportionate reactions among politicians, and the British were prime examples.

To my knowledge there has been only one significant instance in the twentieth century in which fighting broke out over international water issues. That occurred over the period 1964–66 when Syria tried unilaterally to divert the headwaters of the Jordan River in retaliation for Israel's diversion of large amounts of Jordan water through its National Water Carrier to the coastal plains and the Negev. Israel met the attempt with air attacks on the diversion worksites. It can be plausibly argued that these clashes set in motion a series of steps that eventually produced the June War of 1967, pitting Israel against Syria, Jordan, and Egypt.

Conventional conflict can be water-induced. For years Syria, in part owing to its unhappiness with Turkish policy to reduce the flow of the Euphrates before it enters Syria, sponsored the training and passage into Turkey of fighters in the Kurdish Workers Party. Turkey accused Syria of aiding and abetting international terrorists and in the early fall of 1998 massed troops on Syria's borders.

Therefore, to suggest that war is not a likely outcome of water disputes (see Beaumont, 1994) is not to deny the passions that international water quite legitimately arouses. It has always been recognized that water is *casus belli*. If we are deprived of it there is no ready substitute. If we are deprived of it long enough, we die. Fouling wells has been an integral part of warfare throughout history. In the 1950s, when fluoridation of U.S. drinking water was first proposed, many Americans feared that it was a Soviet plot to poison them. Yet using water as a weapon provokes nearly the same censure and instinctive abhorrence as does chemical or biological warfare. It will be the weapon of the truly depraved but not a component of a conventional arsenal. As we shall explore in greater detail in the coming pages, it is, from an engineering point of view, very difficult to deprive large populations of large amounts of water. It may be very difficult, not to mention prohibitively costly, to store river water. Unless the party that carries out the storage has use for the water stored, sooner or later it will have to release water downstream or put the storage structure at risk.

There will be conflict, maneuvering, acrimony, and nonviolent retribution surrounding water issues. The real battlefields for developing countries are likely to be found inside international institutions such specialized UN agencies, the World Bank, and the regional development banks. Here contesting riparians will jockey for position, line up support, seek to place their people in key positions, and try to mute the voices of their rivals. In the 1980s, for example, 80 percent of World Bank lending to the agricultural sectors of the developing world was devoted to irrigation infrastructure. Foreign assistance to irrigation in general averaged $2 billion a year from 1980 on (FAO, World Bank, and UNDP, 1995:11; Briscoe, 1999a and 1999b).

So, too, we should expect intense lobbying of large private firms in the water business. Turkey's Southeast Anatolia project, when it is finished in ten to fifteen years, will have cost $30 billion; a similar amount is projected for China's

Three Gorges project on the Yangtze. Egypt's New Valley scheme is projected to mobilize $90 billion over thirty years. The contracts spun off of such big projects are of a magnitude to win the sponsoring countries high levels of support from large private firms that carry great weight with their own governments.

In this kind of warfare, the battles are not open to the public. Private firms with large potential stakes in yet to be awarded contracts lobby their legislatures to approve aid packages, export credits, and the sale of advanced technologies and equipment to developing countries to help finance multibillion-dollar water development projects. Legislators themselves see clear advantages for their local manufacturing and exporting firms in nailing down deals.[3] The better-off developing countries usually win these battles as they launch the largest projects and import the most goods. They may also be more skilled. Among the Nile riparians, the Egyptians have carefully built a formidable position in several multilateral and bilateral aid and credit institutions. Their positions are dug-in, and to their adversaries seemingly unassailable. The weaker and the poorer engage in guerrilla tactics, trying to outflank the dug-in positions or to attack by night. Occasionally they invest in the time-consuming and often frustrating effort to forge coalitions capable of assaulting the entrenched positions head on.

It does not require the possibility of armed conflict to make international water issues serious. Water constitutes an ever larger constraint to nearly all forms of human activity and concern from agricultural production to hygiene, leisure, environmental protection, and aesthetics. Even with the best of will toward our neighbors, we tend to use water inefficiently. It is not a question of whether we can afford profligacy shaped by habits as old as humankind; rather, it is a question of the modalities of becoming more efficient and of what interests will be gored as we do so. There will be pain enough without war. Inefficient and destructive unilateralism on the part of "sovereign users" will surely continue, but it will become increasingly incompatible with maximizing general welfare. The guns need never fire, no less fall silent, for this to be a compelling story.

The story is also about a resource and its management that ranks with steel plants and nuclear tests in the iconography of the development process. The miles of rectilinear irrigation canals dissecting the landscape, the high-tension wires marching toward the distant horizon, the mighty concrete shells with roaring plumes of water bursting from sluice ways are as seductive to leaders and the led as military parades or rocket launches. They have the advantage of being more palatable to the donors. Warren Ilchman once qualified these giant projects as reflecting leaders' "Edifice Complex." James Scott has depicted them as the embodiment of what he calls "high modernism . . . best conceived as a strong (one might even say muscle-bound) version of the beliefs in scientific and technical progress that were associated with industrialization in Western Europe and North America from roughly 1830 until World War I" (Scott, 1998:89). As we shall see, David Lilienthal (1953), the driving force behind the Tennessee Valley Authority, espoused high modernism well after World War II as did Point Four, USAID, the World Bank, and most UN specialized agencies. The invocation of high modernism justified any amount of unilateralism, and continues to do so today. Solving collective action problems in the Nile basin requires a more humble, less ambitious, and more patient philosophy on the part of the concerned parties than has so far been manifested. Hubris and cooperation make awkward bedfellows.

I believe that the so far unsuccessful quest for cooperation in the Nile basin can yield empirical tests for many of our theoretic understandings of collective action and regime formation. It also bears lessons for other complex transboundary rivers. I am particularly concerned by basins in which agriculture still plays a major role and where the management of surface water resources is of paramount importance. The following list is indicative and by no means exhaustive:[4]

The Semi-Arid Tropics

• The Indus: shared by India and Pakistan; bilateral accord.
• The Ganges-Brahmaputra: shared by India, Bangladesh, and Nepal; bilateral accord between India and Bangladesh.
• The Tigris-Euphrates: shared by Turkey, Syria, and Iraq; two bilateral accords, Turkey-Syria and Syria-Iraq.
• The Jordan: shared by Israel, Jordan, Syria, and the Palestinian Authority; bilateral accord between Israel and Jordan.
• The Nile: shared by Egypt, Sudan, Ethiopia, Eritrea, Kenya, Tanzania, Burundi, Rwanda, Congo, and Uganda. Egypt and Sudan are joined by a water-sharing accord, while Egypt and Uganda are bound by an accord for the operation of the Owen Falls Dam.
• The Niger: shared by Mail, Nigeria, Niger, Algeria, Guinea (source), Cameroon, Burkina Faso, Benin, Ivory Coast, Chad; multilateral accord.
• The Senegal: shared by Senegal, Mali, and Mauritania; trilateral accord.
• The Zambezi: shared by Zambia, Angola, Zimbabwe, Mozambique, Malawi,

Botswana, Tanzania, and Namibia; one bilateral accord between Zambia and Zimbabwe.
- The Colorado and the Rio Grande: shared by the United States and Mexico; two bilateral accords.

It is in the semi-arid tropics that surface irrigation is most widely practiced and where competition in transboundary basins is likely to be most intense.

Colder Semi-Arid

- The Syr Darya and the Amu Darya: shared by Kyrgystan, Kazakhstan, Turkmenistan, Tajikistan, and Uzbekistan; search for accords under way.

These two Central Asian river systems differ from the rivers in the semi-arid tropics only with respect to their much lower winter temperatures. Competition for scarce water in these two basins is already intense, particularly in the lower reaches dominated by Uzbekistan and Turkmenistan.

The Wet(ter) Tropics

- The Mekong: shared by China, Cambodia, Laos, Vietnam, and Thailand; multilateral accord among Cambodia, Laos, Thailand, and Viet Nam.
- The Congo: shared by Congo, Central African Republic, Angola, Zambia, Tanzania, Cameroon, Burundi, and Rwanda; no accord.
- The Amazon: shared by Peru, Ecuador, Colombia, and Brazil; no current accord.
- The Plate (La Plata): shared by Brazil, Argentina, Paraguay, Uruguay, and Bolivia; multilateral accord.

In some systems in the wet(ter) tropics irrigated agriculture is intensively practiced, but the availability of substantial rainfall eases competition for water. Nonetheless, rainfall always fluctuates seasonally so that periodic shortages are common.

Temperate Zone

- The Danube: shared by Romania, Yugoslavia, Hungary, Austria, Czech Republic, Germany, Slovakia, Bulgaria, Russia, Switzerland, Italy, Poland, Albania; several bilateral and multilateral accords.
- The Rhine: shared by Germany, Switzerland, France, Netherlands, Austria, France, the Netherlands, Luxembourg, Belgium, and Lichtenstein; several bilateral and multilateral accords.

• The Columbia and the Great Lakes: shared by the United States and Canada; bilateral accords.

In these systems conflict, if any, is likely to be about pollution of the resource and environmental damage, such as destruction of wetlands. Cooperation, if any, is most likely to be achieved with respect to navigation and flood control.

The world's surface water, and, indeed, its aquifers, are under increasing stress, regardless of zone or ecological niche. Virtually all the nations of the world recognize the problem, and most advocate some form of collective action. What the example of the Nile basin may tell us is in what areas cooperation is most (or least) likely, how much can be done in the domestic arena to prepare for cooperation, what priority developing nations may place on cooperative solutions to transboundary water problems, and how much leverage the donor community can exert in the search for cooperative regimes.

Chapter 1 Collective Action
and the Search for a Regime

WHERE DOES COLLECTIVE ACTION
COME FROM?

The empirical puzzle that I examine in this book is that of ten, nominally sovereign, states in the Nile basin that share, to widely varying degrees, the water resources of the basin but that have not developed any comprehensive set of rules and understandings that regulate that sharing. These are, in the terms of water law, the riparian states and will be referred to henceforth as the riparians. They thus face a collective action problem, but there is no consensus among them on how much of a problem it really is. Moreover, the water resources themselves have been only partially appropriated by the riparians through use rights. Only two of the riparians (Egypt and the Sudan) have given formal, reciprocal acknowledgment of their rights, while the other eight riparians recognize neither the Egypto-Sudanese claims nor any other riparian claims. All the riparians periodically extol the virtues of cooperation and coordination in water use, and international funding agencies assure the riparians that cooperation will yield important net benefits, albeit unevenly. How likely is it that we will see enhanced cooperation in the Nile basin, and what best explains progress, or its absence, toward greater cooperation among the Nile riparians?

Those questions reflect merely some of the enduring puzzles with which political economists and virtually all social scientists have grappled across the centuries. The master puzzle is the collective action problem. How is it that a set of actors with divergent interests coordinate their actions to achieve common purposes in those areas where their interests converge? A derivative puzzle is the means by which this coordination is brought about and sustained *voluntarily*. From that puzzle comes the third: how do institutions that foster voluntary cooperation develop and endure? Finally, we face the puzzle of how nominally sovereign actors voluntarily establish procedures jointly to manage and exploit common pool and common property resources. All these terms require more detailed elaboration.

The standard explanations of collective action, or its absence, tend to fall within four explanatory frameworks:

Community, whereby the norms and expectations of a defined group exert such pressure on any individual in that group that she or he "voluntarily" contributes to the public good. Emile Durkheim is the standard bearer of this understanding.

Market, whereby the rational and uncoordinated actions of individuals contribute to the provision of the public good. Adam Smith is the foremost proponent of this school of thought.

Contract, whereby rational actors negotiate an agreement that yields the highest returns for all parties to the contract. John Locke is the leading theorist of contractual solutions to collective action problems.

Hierarchy, whereby solutions to collective action problems are imposed and enforced by dominant powers for whom the benefits of collective action are greater than the costs. Thomas Hobbes is most closely associated with hierarchical solutions (Lichbach, 1996:22–25).

In all forms of economic, social, and political life those who seek to promote collective action simultaneously try to devise formulae to reduce monitoring and its attendant transaction costs and to increase voluntary compliance. Communities may be of the organic variety, emerging slowly and spontaneously, held together by religion, blood, or patriotism. Such groups act collectively because members fear loss of respect if they shirk group duties and gain prestige if they perform well. For some the fear of God is sufficient motivation to adhere to group goals, whereas for others fear of ostracism and expulsion is equally effective.

Other communities are formed by compact and contract. A necessary condition is that each member shares, nominally, in the stated goals of the community. Although membership is voluntary, each member must nonetheless follow the stated procedures of the organization, support it through dues, and risk expulsion if she fails in either respect. Such groups may range from a high school stamp club to the American Medical Association to the North Atlantic Treaty Organization.

Collective action theory, as propounded by Mancur Olson (1971) and others, is better at explaining noncooperation (defection) than it is at explaining collective action itself. It is logical that if an individual can benefit from the provision of a public good, let us say cooperative action that produces net benefits for the participants, without sharing in the costs, then she will free-ride. If defection is not effectively sanctioned, more participants will free-ride until the public good is no longer provided.

Yet we know that collective action is not rare. It is difficult to initiate and difficult to sustain, but solutions have been found. While the explanation of defection and free riding is simple and powerful, the formal explanations for collective action are logically unsatisfactory, or, as Lichbach puts it, "incomplete" (1996:207). For example, contracted cooperation in the absence of a hierarchical authority may not be enforceable or sustainable. Markets that function with low transaction costs may require communities to monitor them. Hegemonic powers may likewise need communities in order to reduce the costs of authoritarian solutions to collective action problems. No single variable offers a complete, logical explanation of collective action. Moreover, each explanation contains a kind of infinite regress within it. Communitarian solutions, for instance, raise the question of the origins of the community itself. Each explanation tends to presuppose a prior successful collective action that itself needs explaining.

Cooperation and coordination bear costs to the beneficiaries of collective action, and because different actors benefit differently from cooperation, some will bear heavier costs than others. Douglass North (1990) and many others have emphasized the role of institutions in eliciting voluntary compliance, reducing monitoring costs, and evening out the costs of collective action. North's understanding of institutions is conceptually broad and includes formal rules and explicit norms, informal understandings, and expectations about others' behavior based on experience. It is important to note that voluntary and coercive solutions are relative concepts. Even the most voluntaristic solutions require monitoring and the application of sanctions against defectors.

Well-functioning institutions provide a framework for relatively voluntary solutions to collective action problems sustained over time. In that sense there is little to distinguish institutional solutions from the establishment of a regime. A regime consists in "sets of principles, norms, rules and decision-making procedures around which actors' expectations converge" (Krasner, 1983b:2). However, regimes tend to come about through the determined actions of a preponderant power, sometimes referred to as a hegemon, that imposes institutional solutions. Those parties to a collective enterprise who stand to benefit the most may have an incentive to enforce adherence to collective goals (see Olson, 1971:30–31; Hardin, 1982:73–75).

An imposed or hierarchical solution does not rely on voluntary compliance and will entail relatively high monitoring costs for the power that establishes the regime. The monitoring is necessary to prevent cheating or outright defection on the part of the weaker and less motivated parties to the collective regime. These costs may outweigh the benefits of the collective action to the dominant power. Classic examples are landlords who have to employ many armed retainers to make sure sharecroppers deliver their contracted shares of total produce; the cost of the retainers may exceed the value of the additional produce obtained by monitoring. A country may spend more on employing inspectors to pursue tax dodgers than the net gain in tax revenues resulting from the pursuit. Commercial contracts may be enforceable only through lengthy legal investigation and court action, thereby nullifying the profits the contract was intended to generate.

Those who carry out transactions within a mature set of institutionalized arrangements know what to expect, how to conduct themselves, and, whatever the relative degrees of coercion and voluntary compliance, most actors can assume that most other actors will uphold their obligations. Therefore, those concerned with solutions to specific collective action problems should concentrate on institutional design (see Ostrom, 1992). That, in turn, will entail far more than rules, sanctions, and formal organizations, although they will be necessary, but in addition a system of incentives and payoffs that will make the collective action sustainable over time. When a collective action problem is solved over time because of a recognized system of institutional arrangements and incentives (and disincentives), then we may speak of the establishment of a regime.

The creation of institutions is itself a collective action problem (Bates, 1988), and, in parallel fashion, there is no logically complete explanation of the emergence of institutions and regimes. The process of institution and regime formation is sui generis and often based on post hoc analyses. With hindsight we can explain the process, but we may not know of all the would-be institutions and regimes that never made it. We can learn from historical processes lessons that we can then apply in consciously trying to craft new regimes. Note that if there is a "we" trying to craft institutions and regimes, this creates another collective action begging for explanation.

In one of the most sophisticated attempts to answer this question, Robert Bates has analyzed the creation of the International Coffee Agreement and the International Coffee Organization (ICO), 1962–89 (see Bates, 1997). The collective action problem was to coordinate the international marketing strategies of the major coffee producer-exporters, primarily Brazil and Colombia, so as to maximize their export earnings. Part of the challenge was to deny the "public good" of higher prices to producing nations that ignored or did not adhere to the production quotas set up by the agreement. The solution to the collective action problem could not be achieved by the major producers alone; the only effective weapon at their disposal against defection was to dump coffee on international markets, thereby driving down prices in an effort to punish those who marketed above quota. But that weapon was two-edged because the major producers would themselves experience substantial losses in export revenues. Consequently, another set of actors, the consumers, had to be brought into the solution in order for the cartel to work. The consumers were represented by the large roasting firms that bought the coffee and marketed it in the major consuming nations. These firms were strategically placed to police producers, refusing to buy coffee from those who did not adhere to quotas (p. 152).

The final piece to the solution was the willingness of the U.S. administration and Congress to accept higher prices for U.S. consumers of coffee than would have prevailed in the absence of an effective cartel. This willingness stemmed from Cold War concerns to stabilize the economies of the poor nations of Latin America which were exposed, so it was feared, to the revolutionary winds blowing from Cuba. U.S. support for the ICO was the functional equivalent of financial aid, but one that did not require annual appropriations and which was mainly invisible to taxpayers and to consumers. It is no surprise that the ICO collapsed at the end of the Cold War.

To answer the question where did this institution come from, Bates stresses the interaction of evolving economic interests and institutions: "Changes in the domestic political structure of Colombia and Brazil freed their governments to collude at the international level. Changes in the structure of political institutions strengthened their governments' capacity to capture, subdue, and

prey upon their coffee industries. Interest groups thus do not determine policies. Rather, policies are the result of political processes in which groups gain,
or lose, power, *depending upon the structure of domestic political institutions*"
(Bates, 1997:119, emphasis added). We may conclude that international institutions come from a complex, interactive process rooted in domestic interests
and institutions. In the instance of the ICO, we know what happened and more
or less why and how. But if we were to ask what regime, if any, is likely to
emerge after the collapse of the ICO, could we find a parsimonious, theoretically
satisfactory answer in what we know of existing economic interests and institutions in the major coffee-producing nations, and in the nature of the post–
Cold War world?

The collapse of regimes and institutions, and what replaces them, is a facet
of the master puzzle particularly relevant to this study. (The Nile basin was
once governed by a hegemonic authority, Great Britain, which imposed a
regime, but I will hold that discussion for Chapter 3.) Another recent example
of a collapsed hydrological regime is to be found in Central Asia. When Nikita
Khrushchev identified the Central Asian republics of the USSR as the locus for
the Soviet Union's agricultural future, Moscow took over planning the use of
water resources in the two major river systems of the region, the Amu Darya
and the Syr Darya. The major focus of agricultural expansion through irrigation was in Uzbekistan and secondarily in Turkmenistan. The three upstream
republics of Kazakhstan, Kyrgyzstan, and Tajikistan were to be the sites of
water storage, hydropower, and flood control works. Agricultural value-
added would accrue mainly to Uzbekistan and Turkmenistan. The regional
division of labor and water was authoritatively handed down by Moscow despite the theoretic autonomy of the five republics. This was a hegemonic solution to regime maintenance. There was no question of voluntary compliance
as Moscow controlled the investment resources, the party, and coercive force.
It is now the case that all five republics are independent, sovereign states. They
are relatively free to define their national interests. They have no experience of
negotiating with one another over water issues (see Micklin, 1991; O'Hara,
2000:429). The hegemonic regime has been replaced by no regime at all (although the World Bank has been trying to help create one; see Boisson de Chazournes, 1998).

As we shall see in Chapter 3, two regimes, the British and the Cold War, have
collapsed in the Nile basin. Will it require a hegemon to impose a new regime,
or is a more voluntaristic process possible?

There is much empirical evidence to sustain the proposition that voluntary compliance with international institutions and regimes is not only possible, it is already the norm, Scholars such as Michael Taylor (1990) and Abram Chayes and Antonia Chayes (1995) rightly point out that the "realist school" has advanced a tautology when it posits that a state will respect a treaty only when it is in its interest to do so (1995:3).

The Chayeses' focus is on the elaboration of international regimes to manage the environment (see also Young, 1994). They want to estimate the chances of voluntary compliance to the regulations embodied in international conventions on the release of chlorofluorocarbons (CFCs) that threaten the ozone layer, the emission of greenhouse gases, protection of whale populations, bans on trade in endangered species, and so on. Such conventions present collective action problems of staggering proportions because the potential parties are most or all of the world's states, benefits and costs will be unevenly distributed, and the public good provided (less global warming, protection of the ozone layer, etc.) will be available even to those who bear none of the costs. Still, the Chayeses are relatively optimistic. They examine the history of a number of international treaties and joint undertakings, including arms control and inspection, the Barcelona Convention of 1976 to limit land-based pollution of the Mediterranean Sea, the application through the International Labor Organization of norms and standards in labor practices, and the unusual instances in which international regimes actually resort to sanctions to monitor the behavior of defectors (the case of the UN vis-à-vis South Africa under apartheid or toward Iraq when it occupied Kuwait).

They conclude from their review of past experience that most nations most of the time honor treaty obligations. The reasons for this general, dominant tendency to comply are manifold. States invest a great deal of time and expertise in negotiating treaties and accords. The negotiating process itself creates constituencies for maintenance and compliance; these are not only national but also transnational groups of experts in the treaty domain who share a set of concerns, norms, a "language," and sometimes personal friendships. Peter Haas (1990, 1993), in his study of the Mediterranean regime, called these epistemic communities. Further, treaty compliance confers international legitimacy and defection draws opprobrium or worse. In this vein the Chayeses point out that even when a state tries to evade treaty obligations, it will seldom do so brazenly but will instead seek legalistic interpretations of the rules to provide it "wiggle room." The appearance of compliance is important (1995:13).

They question a possible rejoinder—that regimes are sustained by the threat of sanctions—noting that few regimes have effective sanction mechanisms and that where there is provision for sanctions they are seldom applied. Severe sanctions, up to and including expulsion from the regime, are potentially self-defeating; only by keeping parties in the regime will it be effective (1995:87). The application of sanctions, I would note, also presents a classic collective action problem: the costs of application will be unevenly distributed but the benefits (if any) can be denied to no one. The temptation to defect (not to join in trade sanctions, for example) will be great.

The authors stress that international law provides the common discourse of international regimes (1995:126). This language is constantly reinforced through the activities that sustain the regime. These same activities are the currency of compliance:

1. data-gathering and regular reporting;
2. disclosure and notification (country X plans to build a paper mill on an internationally regulated river);
3. dissemination of information to the regime; administration, national constituents, and the international community;
4. training personnel expert in the technical, legal, and policy aspects of the regime; and
5. verification of compliance and monitoring of required actions under the regime.

It is probably true that with respect to most international treaties and conventions, compliance is the norm and defection the exception, but part of the explanation for this may lie in the fact that most treaties and accords are not very demanding, compliance is hard to monitor, and there are no credible sanctions against defection. Adherence is relatively cheap; it buys observer status at a minimum, and access to the data and reporting of other members. As the Chayeses stress, it does confer some legitimacy in the eyes of the international community. In contrast, where bottom-line national interests are at stake, particularly in the military and security domains, the creation of a regime in the first place, and its maintenance thereafter, may not obey the logic put forth by the Chayeses. India and Pakistan threw international legitimacy to the winds in 1998 when they both openly flouted the nuclear nonproliferation regime backed by the United States. In a very different arena, the United Kingdom, in 1992, pulled out of the European Monetary System that regulates exchange rate policy among EC members, when the domestic costs of defending the pound sterling became unsupportable. In 1995, after an assassination attempt on the life of Egyptian President Hosni Mubarak, Egypt accused the Sudan of being behind the attempt. The Permanent Joint Technical Committee on the Nile, a bilateral body set up in 1959 to monitor the agreement between the two countries on the utilization of the Nile, missed some of its quarterly meetings.

The control and use of surface water and groundwater are, for many states, a matter of national security. An economy cannot be retooled to do without water. In many nations water is essential to agriculture, power generation, and transportation. For such countries the security stakes implicit in supranational regimes are very high. On the one hand, the possibility that an international regime can introduce credible assurances about future supply, and its quality, is a powerful incentive to join, but on the other hand, joining may well entail giving up something judged to be of national interest.

THE NATURE OF THE RESOURCE

The structuring of a regime to manage natural resources or commodity markets will be partially determined by the nature of the resource itself. The ICO regime described by Bates was based on a storable, nonperishable commodity produced in a geographically limited number of countries. Water in international watercourses is fundamentally different.

- Storage. Unlike coffee, oil, or gold, water is difficult and costly to store. The former are stocks, whereas the latter, transboundary water, is quite literally a flow.[1] Stocking it is potentially dangerous to the stocker. Prolonged nonrelease of water will tax the storage infrastructure itself, putting at risk one's own population downstream of the facility. Saudi Arabia can keep its petroleum in the ground, but an upstream state like Ethiopia cannot keep its water in the highlands. The only way that supply to downstream riparians can be threatened is by developing the infrastructure and the agricultural base to use the water in-house as it were. This is what Turkey is doing in the upper reaches of the Euphrates River (see below). Such infrastructure, however, normally costs in the billions of dollars.
- Cartelization. Traded commodities can be managed by a cartel of producers and/or suppliers that hold a stock and jointly manage it. Water can be marketed, but as we shall see below, it is still a fairly rare occurrence and nearly unheard of internationally.
- Sequential Use. The use of water in a transboundary watercourse is sequential, not joint or simultaneous. Therefore, any cartel of suppliers is likely to emerge only among a subset of upstream riparians.

• Reaction of Consumers. Again, unlike coffee, gold, or petroleum, if water is stored and withheld by a supplier, that action is likely to be *casus belli*. There are no substitutes for fresh water except other sources of surface and/or groundwater. There are minimum amounts (Falkenmark, 1989, and Glieck, 1996 have suggested 100 liters per capita per day) that are vital for survival and basic hygiene. This amount, or something like it, cannot be foregone nor replaced with anything else. The elasticities of demand for water for human consumption are, thus, low. The possibilities for demand management for nonhuman use, especially in agriculture, which typically accounts for 80 percent of such use,[2] are very high.

• Localization or Regionalization. Changes in the quality or nature of the water in an international watercourse will affect only the riparians themselves (although coastal pollution through river runoff and effluents may affect nonriparians). The downstream riparians are likely to receive the cumulative effects of (mis)use practices upstream. The confinement of environmental effects of changing water quality and quantity in international watercourses to the basin itself is one of the few factors that would facilitate the establishment of a use regime.

It is important to point out that international watercourses, including the Nile, do not constitute common *pool* resources that can be exploited jointly and simultaneously by the riparians in the basin. They are, rather, common *property* resources. Only the riparians can use them (exclusiveness), but among the riparians themselves exclusion is difficult. There are rivalries in use as well; the supply of the good is limited. In both senses transboundary waters are "impure" public goods (Ostrom, 1991:31–32).

THE UNIT OF ANALYSIS

Economists and environmentalists are likely to see the basin or watershed as the proper unit of analysis in approaching questions of water management in transboundary watercourses. The use of water by one riparian will produce externalities, both negative and positive, the costs or benefits of which will be borne by other riparians. A negative externality would be the dumping of industrial wastes upstream so that a downstream riparian would have to pay for the cleanup or for the health effects on its population. A positive externality might arise from the construction of a dam and power station upstream that afforded protection against floods downstream. From an economic point of view, treating the basin as a unit means that these externalities are internalized. By so doing more rational decisions can be made about the opportunity costs of investments in particular projects, about the most efficient use of a scarce water resource, and about ways to maximize welfare for all who inhabit the basin (Rogers, 1991; Marty, 1997:18–19). For environmentalists, the basin likewise is the proper unit of analysis as changes, natural or human-made, in one part of the basin will have ramifications throughout the basin. The goal for environmentalists is to protect to the extent possible the natural regime of the basin, and to ensure that human effects are rendered the least harmful.

International relations experts are likely to view the state as the proper unit of analysis. However severe the constraints upon it, it is still the state in the international arena that is the locus of sovereign decision-making. Only by understanding what drives state policy will we be able to anticipate what it will take to bring about a regime.

International river basins are large and complex, often with many sovereign players. Therefore it may be at the sub-basin level that we find the proper unit of analysis. At that level there are likely to be fewer players, the asymmetries of costs and benefits somewhat less, and the coordination challenge simplified. Sub-basin institutions and regimes may be the stepping stones to basin-wide understandings. At the same time they run the risk of intensifying intrabasin rivalries if clusters of riparians strike water-sharing deals at the expense of the others.[3]

THE KEY PLAYERS

Insofar as the Nile basin is concerned, three types of actors will be crucial to producing the institutions and practices that undergird a regime. First are the executive authorities and policy-making elites of the riparian states. No domestic, civilian constituencies have yet developed sufficient weight or definition of interests to influence national policy. It will be very difficult for them to do so under any circumstances because transboundary water is seen as a matter of national security and strategic interest. National policy elites will not willingly permit open debate of these issues in the domestic political arena.

Second are the international donor institutions, primary among them the World Bank but also including the specialized UN agencies (UNDP, FAO, WMO, and UNEP), and the regional development banks. These institutions have the financial resources to cover the start-up costs of basin-wide cooperation, including the development of jointly managed infrastructure. It is hard for them to move too far ahead of the perceived national interests of the states that are their official clients, but they have shown some inclination to promote supranational solutions to collective action problems. In this sense, donor institutions can prepare a middle ground between the hegemonic imposition of a regime and voluntary compliance. These institutions can induce cooperation through financial leverage by providing the material incentives to compensate reluctant participants for the costs, perceived or real, of participation.

International donor institutions are not, however, neutral. Their policies and priorities reflect an unstable balance of the professional preferences of the experts who staff them with the national objectives of their member states. Their analyses and recommendations are the product of bargaining and compromise among these constituents, and the richer contributing nations can steer their interventions. In many respects they are subarenas for the playing out of the same rivalries that one finds in the watercourse itself. It is in the corridors and conference rooms of the multilateral donor institutions that the real water wars take place, mercifully without bloodshed.

It is not analytically clear why these third parties are more motivated, persistent, and clear-sighted than the riparians themselves in promoting cooperation. There are certainly institutional and professional rewards that result from success. Being external to some extent to the problem to be solved, they may possess a credibility that the parties most directly involved lack. Finally, they have the resources to move the parties toward cooperation. They may also administer "selective incentives," not directly related to the public good itself, to specific riparians. In short, they may be crucial to initiating and sustaining a movement toward formal cooperation (Lichbach, 1996:171–72, 189).

Multinational firms offering capital goods such as turbines, pumps, regulators, power transmission equipment, as well as engineering, construction, and contracting services, constitute the third set of actors. Water development in all its facets offers multibillion-dollar markets to these firms. The scales and capital-intensity with which they prefer to deal often mesh nicely with the preferences of national elites for grandiose development projects that also spin off side payments to the officials of contracting agencies. As large-scale water management projects have lost favor in the more developed nations, giant firms like Mitsubishi, General Electric, Siemens, Bechtel, and Asea, Brown, Bovery are in the hunt for third-world markets for their services and equipment. They are probably indifferent as to whether projects are carried out under national auspices, such as Turkey's $30 billion Southeast Anatolian project, or transnationally, such as the movement of natural gas, petroleum, or electricity.

It should be noted that most bilateral aid agencies generally try to promote the interests of firms domiciled in their countries, and that they are also most comfortable dealing with other nation-states. In the Nile basin, for instance, the only multilateral initiative with which USAID has been associated is the Inter-Governmental Authority on Drought and Development (IGADD; see Chapters 6 and 7), which groups some states in the basin and some outside it.

Nongovernmental organizations concerned with environmental issues provide important services that are not always appreciated by their targets. NGOs are important players but not of the same weight as policy elites, donors, and corporations. They use their supporters' funds to pay for data-gathering, monitoring, and the provision of some expertise at the initial stages of regime construction, and thereafter. Because they are whistle-blowers and policy gadflies, their "interference" is often resented, but they have unquestionably obliged local and regional actors to worry about the consequences of their projects and policies. They have done the same with respect to multilateral donors, of which the World Bank has been a favorite target over the years.

We should expect, therefore, fairly complex two-level games with several players (Putnam, 1988). The pivotal players are national policy-making and political elites operating in both the international and domestic arenas (or levels). They will use domestic interests and pressures to help extract better terms at the international or regime level, and, in turn, use pressures generated internationally to sell policies to reluctant domestic constituencies.

LEGAL DISCOURSE

For more than three decades, the General Assembly of the UN, through the auspices of the International Law Commission (ILC), has sought to codify the rules of use of water in transboundary basins. The ILC has produced several drafts of the Law of the Non-Navigational Uses of International Watercourses. Each draft has tended to drift further toward a lowest common denominator until the General Assembly, on May 21, 1997, approved a final text, opened for the adherence of individual member states (UN General Assembly, 1997). Nonetheless, as the Chayeses (1995) have argued, the deliberations of the ILC have provided a discourse for discussion of the structuring of watercourse regimes. The legal expertise mobilized from the member countries has now spanned a generation.

What was finally adopted, amidst widespread abstentions and split votes, was a framework agreement that may yield formal legal norms. Yet it will not be by law that regimes are created and maintained. International law merely provides conceptual tools to help guide negotiations among sovereign actors. If, in any specific river basin, important riparians are not genuinely interested in cooperation, then an international convention will be powerless to constrain their behavior. International law, like all law, embodies centuries of experience and the lessons of trial and error and of costly, destructive conflicts. The corpus of law that nourishes a regime contains the sunk costs of others', often painful, learning. Contemporary actors, groping toward mutual understanding, would be foolish to forego these distilled lessons. There will, however, be no credible sanctions contained in a convention on international watercourses that could prevent a party to it from defecting. In that sense international law can do no more than establish the rules and procedures of voluntary compliance, but that is already a lot.

As in virtually all legal discussion of water rights, the ILC has had to grapple with what might be called the Master Principle of appropriation. Simply stated, it is that whoever uses the water first thereby establishes a claim or right to it. Any use or user coming second in time cannot legally diminish the right that the first user has claimed. The maxim is "first in time, first in right." In water law of the western United States, these are known as "senior rights"; elsewhere they are known as "acquired (through use) rights" based on the principle of "prior appropriation" (see Boggess, et al., 1993:340–41; Huffaker et al., 2000).

In many river basins, the first, sustained use of water took place in the lower reaches where the slope of the watercourse is most gentle, the deposit of deltaic soils for agriculture the most extensive, and the movement of goods the easiest.[4] The upper reaches are more likely to be rugged, somewhat inaccessible, and ill suited to extensive agriculture. In addition, because rainfall or snowmelt was more abundant than in lower reaches, upstream populations tended not to make direct use of the watercourse itself. Acquired rights were asserted and prior appropriation claimed most often by those situated downstream, and, as their territory typically contained the seat of government and legislative authority, they enshrined the Master Principle. In the Nile basin, Egypt is the temple of acquired rights doctrine.[5]

Closely associated with the principle of acquired rights is that of "appreciable or significant harm." This principle protects acquired rights by warning all second-in-time users to avoid any use that might cause harm to those with senior rights. Some jurists argue that *any* harm, no matter how insignificant, nonetheless constitutes harm, while others, recognizing the veto power inherent in this principle, suggest that the harm must be significant, not only measurable but also clearly deleterious (see Bourne, 1992:81; Nanda, 1995:282).

Appreciable harm is, in economic terms, a most peculiar principle. It is territorially based and conceived. It stipulates that one is not permitted to use one's property in such a manner as to cause harm to another's property (Caflisch, 1998:12). In terms of its logic, it is not clear why harm arising from the use of territory or property should be judged more injurious than harm arising in respect to any established pattern of use. When linked to acquired rights, this principle becomes a serious impediment to any technological innovations that would lead to more efficient use of water (Boggess, 1993:353; Huffaker et al., 2000).[6] Rights are based on use, and if improved use practices lead to a lower water requirement, the acquired right will have to be defined downward. In most other domains the exposure to threat in established practices is considered to be a major stimulus for innovation and competition. When OPEC engineered a dramatic increase in international petroleum prices in 1973, advanced industrial economies experienced significant harm and disruption to established energy use practices, but there were no legal grounds on which to restrain OPEC's behavior (other than to charge the organization with cartelization of the oil market in contravention of GATT rules). All that the consuming nations could do was change their use patterns of petroleum products by reducing consumption and developing technologies to use petroleum more efficiently. Dynamic economies are littered with the remains of victims of appreciable harm—from trolley tracks and barge canals to the steel mills of the rust belt. The sunk costs in obsolete infrastructure, capital equipment, and techniques (the instruments of use) cannot be invoked to defend the status quo.

It could be argued that water is like no other resource. It is vital to life itself, and there are no substitutes for it. But if this were the underlying rationale for the principle of appreciable harm, then we should invoke it in transboundary water disputes only when the actions of a riparian are seen as life-threatening, such as cutting off all water to a given population or contaminating a water supply.

Before we jettison appreciable harm as generally working in favor of the established interests of the most developed riparians, however, let us keep in mind that it could protect the interests of the least developed and of the environment. If use confers rights, then, for example, the swamps in the southern Sudan that are the source of livelihood for the fishing populations there should be maintained as they are. One could also argue for nature's acquired use rights in order to prevent drainage of wetlands and the deforestation of watersheds.

Although the legal point appears to have been lost in contemporary discourse, acquired or senior rights were never conceived of as unqualified. Elwood Mead, writing in 1903, observed: "While the early appropriators were entitled to protection in their use of water, the later comers had an *equal* claim to protection from *enlargement* of those uses" (as cited in Gould, 1988, in turn cited in Huffaker, 2000:266; emphasis added). This principle applies even if the early appropriators lie downstream and are seen, at first blush, as geographically incapable of causing harm to upstream users.

Against the linked principles of acquired rights and appreciable harm has come that of unlimited territorial sovereignty. This principle states that any transboundary resource arising on or traversing a given territory or jurisdiction can be used within that territory as its authorities see fit, regardless of the uses, established or otherwise, of other riparians. This principle was forcefully advanced by U.S. Attorney General Judson Harmon in a dispute with Mexico over the Rio Grande in 1895. Mexico argued that its existing uses of the river's waters were jeopardized by uses upstream in the United States. Harmon rejected the Mexican claims on the basis of unlimited territorial sovereignty. He nonetheless conceded, "out of comity" as he put it, some water to Mexico. This was a grant, in the spirit of good neighborliness, not a recognition of any legal claim Mexico had put forth. From then on, the principle of unlimited territorial sovereignty was referred to as the Harmon Doctrine. In the Nile basin, the current leaders of Ethiopia, from whose territory the sources of the Nile's major tributaries arise, are disciples in all but name of Judson Harmon.

Since 1966, when the so-called Helsinki Rules were drawn up to define a users' code for transboundary watercourses, the international law community has sought to establish procedural norms lying between prior appropriation and unlimited territorial sovereignty. The norms advanced come under the rubric of "equitable use" and the "community of interests" of all riparians. The stress here is on establishing a use regime that takes into account the interests of all riparians and seeks to reconcile rival claims in a way that makes everyone better off (that is, that the value of the public good created through equitable use will outweigh the sum of the private losses incurred by any riparian through participation in the regime).

The framework for equitable use has been elaborated in articles 5 and 6 of the ILC Convention.[7] Article 6 enumerates factors that should be taken into consideration in determining equitable use. They are distillations of the principles laid down in the Helsinki Rules (see Waterbury, 1977) and include:

• geographic, hydrographic, climatological, ecological, and other natural characteristics;
• social and economic needs of the watercourse states;
• dependency of the population upon the water resources;
• existing and potential uses of the water;
• conservation and protection of the water resources; and
• availability of alternatives of corresponding value.

Some significant signals are sent in these guidelines. Potential use is given equal weight with existing use. A discrete wave to environmental needs is made. Alternatives, typically rainfall, are alternatives only if they are roughly of the same value as the surface water. That said, the draft assigns no relative weights or priority to any of these elements; it merely suggests that they be taken into consideration in regime design. As such the guidelines are not of much operational use. The riparians in any watercourse will simply have to negotiate their way toward an assignation of weights and priorities. As we shall see in the next section, it will be very hard to develop any consensus about what constitutes equity, or about which rival claim is the more compelling.[8]

It is often the case that some of the norms sustaining a regime are in contradiction with one another (Chayes and Chayes, 1995:120). That is true with respect to appreciable harm and equitable use. Albert Utton, arguing for the primacy of equitable use, suggests a way by which the two can be partially reconciled. Equitable use, he argues, should be applied mainly to quantity questions and formulas for allocation. Appreciable harm should be reserved to issues of quality, such as harm to ecosystems (Utton, 1996:636). Of course, the two criteria cannot be neatly separated. The ecosystem of the Aral Sea has suffered catastrophic harm due to both the deterioration of water quality resulting from agricultural wastes and a severe reduction in the quantity of water reaching the sea from the Syr Darya and the Amu Darya rivers. Nonetheless, the suggestion is useful and has been applied in other instances such as the 1935 Trail Smelter Case between Canada and the United States on transboundary emissions and pollution (d'Arge and Kneese, 1980:429).

The gradual marshaling of legal support for equitable use formulas has not been smooth. For some time Stephen McCaffrey, the U.S. rapporteur of the ILC, strongly favored a privileged position for appreciable harm (McCaffrey, 1989), and it was only in drafts after 1991 that the counter-proposition—that even if equitable use caused some riparian appreciable harm, it should still prevail—gained some currency (Flint, 1995:199). In a subsequent article, McCaffrey

(1998:22) opined that the 1997 Convention gives precedence to equitable use over appreciable harm.[9] Bourne, by contrast (1992:69), dismissed serious consideration of potential or future use as purely hypothetical and not to be taken into account.[10] It is a fact that it is more feasible to ascertain harm than to promote equity. Damage can be measured, but fairness is in the eye of the beholder.

The World Bank has steadfastly maintained its commitment to protect acquired rights, enshrined in its Operational Directive 7.50 (see World Bank, 1990; Goldberg, 1995; Krishna, 1998; Annexes 2A–2C, 1998) The general guideline is that the World Bank will fund a project on an international watercourse only if it has been fully explained to the other riparians and their acquiescence sought. It is up to the other riparians to document appreciable harm within six months of notification. If they fail to do so, or if their evidence is unconvincing, the World Bank can seek the assessment of a panel of experts in order to render a final decision. In two instances the World Bank has followed this process and overruled those claiming appreciable harm. The first instance involved the Baardhere Dam on the Juba River that flows from Ethiopia into Somalia; the dam was proposed by the Somalis on their downstream portion of the river. Ethiopia, the upstream riparian, had protested but could not demonstrate that it would suffer from the project. A panel of experts was used to render the final assessment of the claim (Goldberg, 1995:158). After the current Ethiopian republic was established in 1993 on the ruins of the dictatorship of Menguistu Haile Meriam, the new authorities sought World Bank financing of agricultural resettlement projects to absorb some of the demobilized soldiers from the previous regime. The schemes required modest amounts of irrigation water from streams feeding into the Blue Nile (or the Abbay, as it is known in Ethiopia). It took four years for the World Bank to go through all the processes of notification and assessment of downstream effects. Egypt tried to stall the project by repeatedly seeking additional information, but ultimately Egypt could not demonstrate appreciable harm. Despite the outcome, some professionals in the World Bank see the endorsement of the protection of acquired rights as tantamount to an award of veto power that can be used frivolously by those with senior rights.

APPLICATIONS

Let us draw together from the foregoing discussion some of the lessons that apply in the Nile basin. These points will be developed in detail in subsequent chapters.

My starting point is that the main issue in the Nile basin is not so much that some riparians might free-ride on the public good of cooperation, but rather that a number of riparians see little value in the public good itself. Thus the biggest obstacles to collective action in the Nile basin are

indifference: riparians see the benefits but do not value them;
ignorance: riparians do not see the benefits or do not understand them; and
priorities: in choosing where to invest limited expertise and resources, priorities
   other than cooperation dominate.

How, then, should we conceive of the public good that cooperation is to provide in the basin? For Egypt, the Sudan for a time (see Chapters 3 and 6), and Uganda unwillingly (Chapter 7), the public good is maintenance of the status quo, that is, a kind of de facto, basin-wide regime determined by the bilateral 1959 agreement between Egypt and the Sudan. Ideally for Egypt basin-wide cooperation would mean the formal acquiescence of all other riparians to the 1959 regime. De jure acquiescence would be best, but de facto nondefection will do as second best. This result could be obtained through a combination of Egyptian hegemony and indifference among the riparians.

Ethiopia, the only significant rival to Egypt in the basin, is not at all indifferent to the search for collective action. Ethiopia, however, wants a new deal that would negate the status quo. It seeks collective action among the riparians to challenge the status quo and to thwart Egypt. The public good Ethiopia seeks to provide is a new, "equitable" basin-wide regime. This result could be achieved through cooperation among a few riparians coupled with indifference among the rest.

Put another way, a quasi-regime already exists. Three riparians—Egypt, the Sudan, and Uganda—currently support it, whereas the others have periodically rejected it. Despite that rejection, only Ethiopia can be said to have defected from the quasi-regime. By contrast, all riparians say they want a new regime yet only Ethiopia has taken steps to promote a vision that contrasts with that of Egypt, the Sudan, and Uganda. The other six riparians have mainly a wait-and-see attitude.

Any collective action will be contractual in nature as was the bilateral agreement of 1959 that set the terms of the current regime. There is no "community" of Nile riparians, not even one as loose as the European Union. The latter, at any rate, has been able to apply common political and economic norms on its members and on any who aspire to join it. The Nile riparians share neither common norms nor any sense of community. Any compact among them will have to be purely utilitarian which means, in turn, that rational, interest-based explanations of cooperation and noncooperation should work well.

The dynamics of the search for collective action and a regime will be powerfully shaped by two variables: the kind of cooperation that is sought and the nature of the resource that collective action would manage. Recall that the International Coffee Organization's quest was driven by the search for a coordinated marketing strategy among coffee producers. Coffee is not perishable; it can be stored, dumped or destroyed in order to affect market prices and, derivatively, to police defectors from the strategy. By contrast, the collective action problem facing the riparians of the Nile basin is to find a formula by which water is allocated among rival uses within and across the riparian states. There is no regional or international market for the resource. It is not clear how defectors in a water-allocation and quality-control regime can be sanctioned. Unlike coffee, or petroleum reserves, the "owners" of water cannot feasibly hold it. Upstream states may have the technical capacity, usually at great cost, to store some water, but it can be held and added to for only so long before the storage structures themselves are put at risk. Downstream states have no supply leverage over defectors. They have to use other resources to apply sanctions—economic, diplomatic, or military. In this respect we need to complicate further what Bates proposes: in addition to interests and institutions we must add the nature of the resource itself because it influences both interests and institutions.

At the same time, the Nile basin collective action problem hews more closely to some of the theoretic literature Bates finds wanting. The riparians, by and large, have acted as unitary states. Domestic interest groups, outside the foreign policy community, have not been directly involved in the search for a regime. Public opinion is important, and water arouses strong political passions, but public opinion is inchoate and has played no effective role in the development of policy. Only in Egypt might we anticipate the interjection of domestic, agricultural interests into the policy process. But to date, these interests have faced no water supply constraints attributable to the actions of other riparians. Even in Egypt, water policy toward other riparians has been developed within the confines of the foreign ministry and the ministry of public works. Therefore, to the extent that the riparians have elaborated explicit policies regarding the basin's water, we can read them as the products of unitary states pursuing what they conceive to be their national interests. Even if we are able to define those interests, there is no reason to believe that they will be stable. Any cooperative regime, if it is to survive, must be open to periodic renegotiation and adjustment as the cost-benefit analysis of each of the parties to the regime changes.

Chapter 2 Negotiating Regimes

International regime formation, as Bates's study (1997) illustrates, is a contingent process, embodying complex interactions of national and subnational interests, spurred on by concerned entrepreneurs, with collective outcomes as contingent as the initiation of the collective action itself. Unfortunately, from the standpoint of social scientific analysis, it is not a process encompassed by ex ante assumptions and hypotheses. I argue here that the process of regime formation itself—legislating, data-gathering, formal institution-building, and negotiating—can provide momentum, the creation of new institutional interests and expertise, and, occasionally, "tipping" moments that lead to formal cooperation. The process is necessary but not sufficient to bring about collective action. The active involvement of a hegemonic party, as well as of external "facilitators," may well be required to sustain the process and to lead to the establishment of a regime.

The Nile basin, like several other major international watercourses, has registered a significant history of inter-riparian negotiations and of regime formation. While the monographic, case study literature is fairly rich,[1] there has not been much systematic comparative analysis of the negotiating dynamics that have led to the successful establishment of water management accords. Some of the notable exceptions to that statement are David Le Marquand (1977), Bruce Mitchell (1990), Elinor Ostrom (1991), and Frank Marty (1997). Donna Lee and Ariel Dinar (1996) provide a useful review of a dozen formal models of integrated river basin planning and management. Probably the most extensive, albeit preliminary, comparative study of concluded river basin treaties is that of Aaron Wolf (1997). Two of his major conclusions are that the two "master" principles (appreciable harm and equitable use) outlined in the previous chapter were seldom explicitly invoked (so much for discourse), and that the dominant tendency was to protect existing uses and downstream states.

THE VEIL OF IGNORANCE

In *A Theory of Justice,* John Rawls (1971:136–42) proposes that in the face of the unknown or the uncertain, parties seeking a cooperative understanding are likely to, and should, honor principles of equity that will best protect them against future surprises. The logic of this proposition is that if the goal is to construct a regime in which equity for all is most likely to be honored, then bargaining from behind the "veil of ignorance" is desirable. Indeed, between 1956 and 1958, Egypt and the Sudan fitfully negotiated a treaty on "The Full Utilization of the Nile" with very little solid data at their disposal. The treaty hinged on the use of flow data, which were misleadingly or falsely accurate. The lack of information did not prevent the successful ratification and implementation of the agreement, and, as I will argue in several places, it embodied more equity than anything that preceded it (see Chapter 3). The main problem was that it engaged only two of the Nile's then nine riparians.

Nevertheless, conventional wisdom regarding watercourse negotiations is that one can never have too much information, and that the absence of data may prove a serious impediment to agreement. This certainly is the view of experts in the World Bank and of those who have been formal facilitators in negotiations (see Chayes and Chayes, 1995; Bingham et al., 1994; Marty 1997:65). Even if ignorance has a useful role to play, it is simply the case that satellite imagery, sophisticated river gauging, accurate measurement of rain and runoff, and the monitoring of changes in vegetation means that we know a great deal about even the most remote watersheds. Computer simulations allow us to model, forecast, and run scenarios. The problem is that the more the protagonists know about what is at stake, the more clear will be the calculus of costs and benefits. That can lead to entrenched positions rather than flexibility. D'Arge and Kneese (1980) emphasized that common pool and common property resource regimes will require compensatory payment mechanisms, including insurance funds, to offset the losses to perceived national wealth that participation in the regime will impose on a specific party to it. Finally, the call for ever more data can be used as a delaying tactic to avoid getting on with negotiations that may not go one's way or to mask a party's lack of technical expertise. The call is endorsed by the international donor community which is generally willing, if not eager, to pay for more data-gathering.

There are still areas of ignorance that may influence the negotiating process. Surprisingly, perhaps, there is not consensus on some basic matters of fact in some watersheds: we do not know the seasonal sources of flood water in the Brahmaputra; we do not know the precipitation regime over Lake Victoria; we do not know the impact, if any, of the Sudd swamps of the southern Sudan on local rainfall. It is, of course, the future that truly remains veiled. Our models and simulations are based on historical data, yet the ramifications of global warming, the long-term impact of watershed degradation, and the increased incidence of extreme weather events suggest that the future may hold some very big and unpleasant surprises. Anticipating those in the principles of a resource regime may lead to Rawlsian outcomes (cf. Young, 1994:53).

In sum, when the parties to a potential accord perceive the alternative to cooperation to be some form of highly destructive behavior, then an equitable deal may be cut even in the absence of "requisite" information.[2] Where the alternative to cooperation appears acceptable, or where the distribution of costs and benefits resulting from cooperation is not well understood, then more information may be needed. Ignorance may be useful in the *design* phase of regime construction; *implementation,* however, requires detailed information so that projects can be properly assessed, opportunity costs for investments estimated, the natural regime of the watercourse continuously measured, the source and nature of pollutants determined, and compliance monitored.

THE INSTITUTIONAL FIX

Over fifty years ago, David Lilienthal, in his triumphalist study of the Tennessee Valley Authority (TVA), addressed the problem of bridging the planner's focus on the basin as the proper unit of analysis with that of the jurist and the politician on the legal jurisdictions dividing the basin. Just as the TVA provided a planning and management institution for the federal states that share the valley, so too, he suggested, could something he called the public development corporation provide the same services in international watercourses. He stipulated: "The use of the public corporation in itself . . . would have only incidental relation to the TVA idea unless it were the instrument of the basic idea of unified development within a natural region" (Lilienthal, 1953:208). In the succeeding decades such suprajurisdictional and supranational authorities have proliferated around the globe.[3] They have enjoyed varying degrees of success and of autonomy and authoritative decision-making. Referring to experience within the Organization of Economic Cooperation and Development countries, d'Arge and Kneese (1980) advocate "common property resource institutions" to regulate the use of the common property and to internalize externalities caused by the use practices of the participants. At a minimum such organizations will be involved in gathering and exchanging data and offering a venue for the periodic meeting of technical experts from the member riparians. They may also serve as a forum for the adjudication of disputes. Some have a legal personality, can contract loans and dispense funds in their own name, and through majority voting mechanisms (admittedly rare) bind all member states to collective decisions. In the final analysis, however, they must find those solutions to common problems that win the voluntary compliance or acquiescence of the riparians. As complex organizations in their own right, they may engender a culture that will enable them to see beyond the interests of the members and to focus on the watercourse as a whole. In that sense the institution would function as "the social planner" that Wolf and Dinar (1994:82; cf. Lichbach, 1996:189 on the "entrepreneur") advocate for the Jordan basin. If an epistemic community is to emerge, it will be within this institutional framework.

If such institutions are given responsibility for managing shared infrastructure, such as dams, power stations, bridges, roads, railroads, and pipelines, then their leverage over possible defectors will rise significantly, as will the costs of defection itself. This was and is the role of the TVA, but it has rarely been replicated at the international level. Whatever shared facilities fall on a given riparian territory tend to remain under the exclusive control of that riparian. When power lines or a pipeline cross a border, the point of control changes. As an example, the 1959 agreement between the Sudan and Egypt established the Permanent Joint Technical Commission (PJTC). It is empowered to monitor implementation of the accord, to carry out feasibility studies for future joint projects, and to negotiate in its own name for financing. In the event of negotiations with other riparians, it is the PJTC that will lead them. It does not, however, manage the Aswan High Dam nor any of the other infrastructure that was installed in light of the agreement.

Rangeley et al. (1994) compare the performance of several African river basin organizations (RBOs). Success, of which there is not a lot, is a function of the number of players and the simplicity of the agenda. The least successful RBOs have many riparians and a complex set of tasks. These include the Nile with ten riparians and no basin-wide organization, the Niger-Benue with nine, Lake Chad with six, and the Volta with six. The most successful have been simple dyads with focused agendas.[4] The paradigm here is, perhaps, the Lesotho-South Africa treaty of 1986 on the Orange River. The treaty created a Permanent Joint Technical Commission to coordinate the construction and operation of a dam and power station in Lesotho, and the delivery and sale of power to South Africa. Each country maintains a parastatal company to operate its portion of the infrastructure. South Africa contributed to the costs of construction in Lesotho, and the treaty provides a formula for the purchase of power. Despite the fact that only two riparians were involved and that the common purpose was hydropower generation, the treaty took several years to negotiate. A similar example is that of the Zambezi River Authority, linking Zambia and Zimbabwe in the management of the Kariba Dam. The dam is jointly owned, but the power stations belong to the two countries.

Intermediate in terms of success are somewhat more complex organizations such as the Office de Mise-en-Valeur du Vallée du Sénégal (OMVS) or the Kagera Basin Organization (KBO). The OMVS is comprised of Mali, Senegal, Mauritania, and Guinea. They jointly own the Manantali Dam in Mali, but all other works are the property of the country on whose territory they fall. To date the main task of the OMVS has been to engineer a simulated annual flood downstream of the dam and to stimulate recessional agriculture and irrigation in the watercourse. The KBO was first set up among Rwanda, Burundi, and Tanzania. Its main purpose was to finance and construct a dam and power station at Rusumo Falls to sell power mainly to Tanzania. Uganda, after the fall of Idi Amin, joined the KBO. For reasons having little to do with institutional design, the KBO has gone nowhere. Rwanda, Burundi, and Uganda have experienced severe political instability and have been unable to find the funding necessary to implement joint projects. Because the Kagera is the furthest source of the (White) Nile, I will have more to say about the KBO (see Chapter 7).

Rangeley and his coauthors offer this advice to designers of RBOs (1994:17):

"The objectives should be well-focused. A wide, ambitious mandate, extending across non-water related sectors and into areas outside the river basin concerned, should be avoided." *Ceteris paribus* that is certainly true, but where the riparians are numerous, the agenda will almost necessarily be complex and will range beyond water issues. It may be that only by putting other goods on the regime's negotiating table, can the compensatory mechanisms that even out asymmetrical benefits and costs be put into place. The authors themselves note (1994:54) that the KBO's most successful project was in telecommunications.

RBOs can be no stronger or more effective than the domestic institutions and regulatory systems of the riparians through which they must work. Rules and regulations on pollution must be enforceable in domestic courts and enforcement of standards left to domestic agencies. The implementation of treaty obligations may become a domestic political issue and compliance evaded because of internal political considerations. The Great Lakes Organization between the United States and Canada must deal with two Canadian provinces, eight U.S. states, and dozens of municipalities and regulatory agencies on both sides of the border. The noblest ideas of the "social planner" can vanish into this complex of jurisdictions, and even though these institutions are likely to be more competent in countries like Canada and the United States than anywhere in the Nile basin, their functional counterparts are there and jealous of their prerogatives. Guillermo Cano (1986:13) and Frank Marty (1997:93) attribute the lack of success of the Del Plata institutions, grouping five riparian states in the southern cone of Latin America, to the weaknesses and rivalries of the domestic institutions of the states signatory to the Plate Basin Treaty. This observation yields two recommendations. First, those concerned with establishing effective regimes have to look as much at strengthening and coordinating domestic agencies and regulations as they do at the organization of the RBO. Second, the RBO should have a well-focused agenda that minimizes encroachment on existing jurisdictions, and a good deal of decentralization in operations so that it can deal directly and flexibly with specific domestic jurisdictions (Mitchell, 1990:216).

To reiterate a point made in discussing the proper unit of analysis, tackling problems at the sub-basin level will have the effect of reducing the number of players, simplifying the agenda and goals, and minimizing the number of existing jurisdictions that have to coordinate with the sub-basin RBO. I have cited the risk of sub-basin collusion at the expense of the basin as a whole, but building a regime at the sub-basin level could also be a stepping-stone to basin-wide institutions.

ALLOCATION CRITERIA

The creation of regimes bears costs that have to be paid up front in terms of changed behavior and practices, sacrificing some constituency interests, creating administrative agencies, and training experts. The delivery of the public good for which these sacrifices are being made comes later. The distribution of costs will often be debated in terms of equity. The defenders of acquired rights will invoke it as readily and with as much legitimacy as those making the case for alternative uses. All environmental conventions face this monumental challenge. Cooper (1997) demonstrates the difficulty of equitable burden-sharing in containing global emissions of greenhouse gases. Should the rich nations pay more or less because their economies are the most dependent upon fossil fuels? Should large, poor populations be rewarded with lower up-front costs? Should the "dirtiest" emissions be heavily taxed, thereby burdening the large, poor populations of China and India, whose economies run on anthracite, with much of the start-up and long-term maintenance costs of a regime? Cooper rejects as a solution the issuance of and trade in emissions permits because no one would agree on the criteria for the initial allocation.

The Marshall Plan provided $4 billion in 1949–50 to help solve Europe's collective action problem of reconstruction. The plan was a classic example of third-party inducement. But how were the funds to be distributed? Thomas Schelling (1997:10) calls the procedure "multi-lateral reciprocal scrutiny": "Each country prepared detailed national accounts showing consumption, investment, dollar earnings and imports, intra-European trade, specifics like per capita fuel and meat consumption, taxes, and government expenditures—anything that might justify a share of U.S. aid. There was never a formula. There were not even criteria; there were 'considerations.'"

If there was any veil of ignorance it lay in the absence of guidelines on how to assess information, not in the absence of information itself. This example demonstrates the *ad hoc nature* and arbitrariness that characterize the initial stages of regime formation. Some sort of starting point has to be established, some initial allocation of rights, and it must be accepted that the initial solution will not be pretty, but it will be better than the status quo. In 1969 the three Canadian provinces of Alberta, Saskatchewan, and Manitoba signed a Master Agreement on Apportionment of transboundary waters among them. Each province agreed to let half the natural flow of the watercourses pass downstream to its neighbor(s). An RBO, the Prairie Provinces Water Board, was created to monitor the agreement. Two subagreements were signed between the two pairs of upstream and downstream provinces, and a third with the federal government. The 50 percent release figure has the ring of something nicely rounded, simple, and inherently equitable. It had nothing to do with the needs that each of the provinces surely would have argued were vital (see Kennett, 1991:56).

Steven Brichieri-Colombi (1996) has delineated a procedure for assigning rights to the riparians of the Nile basin. He singles out three criteria: (1) the proportion of the riparian's population living in the basin, (2) the riparian's share of the total area of the basin, and (3) the average amount of the water used (abstracted) from the watercourse. Brichieri-Colombi arbitrarily assigns a weight of .33 to each variable in order to illustrate his argument, but he stresses that the weighting system should be the result of negotiations among the riparians. He suspects, as do I, that unless they make some arbitrary assignation, like that of the Prairie Provinces of Canada, they will be locked in interminable debate about equity.

After assigning weights, Brichieri-Colombi converts the criteria into "aquas," or rights to a fixed quantity of water that can be traded. Under any weighting system, the reapportionment of water within the basin would be fairly dramatic, and Egypt would stand to lose the most. The Egyptian delegation to the Fourth Nile 2002 Conference in Kampala, at which the paper was presented, was not amused.

Székely (1990:97) argues for three significantly different criteria: rights based on historic patterns of use (prior appropriation); special needs related to the stage of economic development; and the population dependent upon the water course in each state. Perhaps unfairly I would suggest that the subtext is this. The author is from Mexico, the downstream state on all watercourses shared with the United States, and with senior rights. It is far less developed economically, and it has far more of its citizens dependent upon the Rio Grande and the Colorado than does the United States. The United States might counter with criteria of higher value-added in U.S. agriculture, carefully diluted Harmonism (it is ours, after all), and, by now, fairly strong patterns of historical use.

In 1986 William Baumol developed a formal economic model of "superfairness." It is based on a "core" of overlapping interests shared by adherents to a potential cooperative understanding. Once the core is identified, one has the locus of an equilibrium in which each participant prefers her share to that of any other participant (or, presumably, to her share under the status quo ante). Peter Rogers (1991) applied this model to the three-party collective action problem involving India, Nepal, and Bangladesh in the management of the Ganges-

Brahmaputra (for the case itself, see Crow, 1995). Using variables of land area, cultivated area, populations, population density, potential irrigation, and hydropower potential, Rogers was able to calculate the net benefits in rupees to the three riparians so as to identify formally the core of their mutual interests. It is important to keep in mind that "fairness" here is measured in purely economic terms. While it is unlikely that riparians will be guided exclusively by such concerns, the power of exercises like Rogers's is that it can model and represent a win-win outcome to cooperation.

Let us examine some specific criteria. Agriculture uses more water by far than any other economic activity, and in developing countries agriculture still contributes a large share of GDP. The sector meets basic food needs for poor populations, shields countries against the fluctuations of world commodity markets, and contributes to exports. On equity grounds it seems reasonable to protect the claims made by agriculture on water supplies, and a rough and ready measure of what is needed may be determined from arable or irrigable land. For this measure to be useful, however, one must assume fixed, nonevolving technologies in water use. For instance, if water use is based on irrigating a certain surface using open canal and gravity flow delivery to fields, the estimates of water requirements will be wildly different than if the technology involves drip irrigation delivered through pressurized, piped systems. This is precisely the approach used in the Johnston Plan of 1955 to establish a transnational regime in the Jordan Basin among Israel, Jordan, and Syria. The plan used current acreage and current practice to specify allocations of quantities of flow (see Lowi, 1993; Wolf and Dinar, 1994). The Johnston Plan was never officially implemented, but as a formula for start-up it was probably as good as any.

Article 6 of the ILC 1997 Convention mentions alternatives to the water of the watercourse as a relevant criterion, but the mention stresses that the alternative must be of commensurate value. An oft-cited criterion by interested riparians is per capita surface water resources, including rainfall. For example in the negotiations leading up to the 1959 agreement, the Egyptians argued that the Sudan had a major alternative to Nile water because of its relatively high rainfall (Egypt has no rainfall worth speaking of). Egypt makes a similar argument today with regard to Ethiopia. Rainfall is not a very useful criterion because, above all in the semi-arid tropics, the variations around the mean rainfall are very large, and the runoff may not be captured except at high cost. Rainfall may be most concentrated in regions unsuited to cultivation (such as high mountains or the swamps of the southern Sudan) or it may come at a time when it threatens standing crops. The "commensurate value" notion in the ILC

Convention can be given practical meaning by reference to alternative, *utilizable* water resources, including stored water, mean rainfall in appropriate areas, and economically exploitable groundwater.

In sum, one could envisage a scheme that arbitrarily assigned equal weights to these criteria:

- the proportion of water flowing across a riparian's border to the total discharge of the water course (an absolute measure);
- the proportion of the riparian's total population living in the basin (a relative measure);
- the total amount of irrigable land that could be farmed with watercourse water *without extra-basin transfers* (see below);
- the amount of alternative, utilizable water available in aquifers, regionally appropriate rainfall, and stored water (a deductible, as it were);
- a basic needs per cap allocation to protect life and basic health (an absolute measure); and
- an allocation necessary to protect existing wetland and nature's "use rights" (one hopes, an absolute measure).

These six criteria, equally weighted, could yield an initial allocation. (For a partial application, see Gupta, 2001:68.) That allocation will be made more palatable if it is accompanied by a system of tradable rights such as that proposed by Brichieri-Colombi. Only with that in place will riparians not feel locked into the initial allocation. To this should be added the following: allocations, outside of the absolute amounts needed for human consumption and environmental protection, should not be set in quantities, which has been the norm in most agreements, but rather as proportions of flow (see Ostrom, 1991:107; Rosegrant and Binswanger, 1994:1621). This protects the negotiated regime against future, unpredictable changes in flow. Second, the agreements underlying the regime should be open to periodic review and amendment as the parameters of the six criteria listed above change.

A parsimonious but potentially devastating formula is offered by Peter Beaumont (2000:487). The principle is that an upstream state is entitled to 50 percent of all water *generated* on its territory (rainfall and runoff), as opposed to waters that merely run through the country but originate elsewhere. That takes care of allocating half of all flows. The remaining half would have to be allocated by different criteria, and prior appropriation would be the dominant one. By this formula, Beaumont calculates that in the Nile basin 35.8 billion cubic meters (bcm) would go to Ethiopia, 23.3 bcm to Egypt, and 7.4 bcm to the Sudan (2000:493). In light of this formula, Brichieri-Columbi's, from the Egypt-ian point of view, looks relatively benign.

In some ways, the "cleanest" set of criteria by which to allocate water would rest on various measures of economic maximization and efficiencies. Instead of the supranational "social planner" or basin-wide authority, the market would tell riparians where the water should go. The simplest criterion would be that water should be allocated to those uses that yield the highest economic return per unit of water (for an application, see Becker and Zeitouni, 1998). Put in economic terms, the challenge is clear. Over time the supply of impounded or stored water tends to become inelastic as good storage sites are eliminated. De-mand, of course, increases, and agriculture, industry, human consumption, and instream uses (environmental protection) compete for supply. The logic of market signals to guide allocation among rival sectors and uses is strong (Shatanawi and al-Jayousi, 1995). The very fact that there is so much debate over legal rules indicates the general recognition that there is potential excess demand and insufficient supply. A market-determined price to a scarce factor would certainly help us understand the opportunity costs of using that factor in different ways. We can then choose to ignore these signals, but at least we will have a clearer idea of what we are forgoing.[5] As noted above, there is cultural and normative resistance to treating water as a commodity because of its non-substitutability. Marketing water will never be popular.

For a water market to exist, the means to store and deliver the resource must be in place. The engineering and cost challenges in this respect are daunting. Coffee, cotton, or wheat can be stored in warehouses and silos, gold in vaults, and oil left in the ground, but water requires dams, canals, pipes, and pumps if it is to be held and then marketed. Very few countries can afford that kind of infrastructure, and even if a riparian has the means, it cannot hold the resource indefinitely; it must either release it for domestic use or release it downstream. If it is used domestically for power generation, the water will flow to the down-stream riparians anyway.

California has the infrastructure to make water markets feasible, and in the drought years of the late 1980s water was traded on a fairly extensive scale. The transactions were monitored by water districts, which were empowered by the Bureau of Reclamation. What is most striking is that the largest amount traded in any one year, 1989, was 890,000 acre-feet (approximately 1.1 bcm), or only 3 percent of total water use in the state (Rosegrant, 1995:73). It is highly unlikely that traded water in any watercourse will be more than a small fraction of total water use. The initial allocation among users will in fact tie down the bulk of the water available, but the ability to trade at the margins should provide real and psychological relief against the fears of being locked into a given allocation in perpetuity. Trading in water rights can also provide a mechanism for emergency relief for specific users over short periods of time.

It is also unlikely that there will be bulk international trade in water (mineral water for drinking is another matter). The costs of transportation, in most instances, whether by ship or pipeline, are not less than the costs of desalinating brackish or seawater. So far, plans to deliver fresh water by transboundary pipelines or ship have not been cost-effective. At the same time, as Tony Allan has argued, there is extensive international trade in water embodied in agricultural commodities. Allan calls this "virtual water," and the price of the commodity reflects the price of the water in the region in which the commodity is produced (Allan, 1996).

The principle of allocating water to uses that provide the highest returns will not operate well internationally. If it did, riparians with low returns to water might become *rentier* states, selling water at a profit to neighbors who can generate higher returns. If this logic were followed, for instance, a Palestinian entity might find itself selling any water it did not use for human consumption to Israel, where the returns to a unit of water are significantly higher. In fact, it is likely that the riparians with acquired rights will have developed agricultural sectors that yield higher returns per unit of water than any riparians who came "second in time." That is certainly the case for Egypt vis-à-vis all the other riparians in the basin.

In the Euphrates Basin, Iraq is the senior riparian with rights nearly as ancient as Egypt's in the Nile. Syria developed use rights only in the middle 1970s, accompanied by Turkey, the upstream state in the system. Turkey has begun one of the largest regional development schemes of the twentieth century, the Southeast Anatolia Project (GAP is its Turkish acronym) (Bilen, 1997; Kolars and Mitchell, 1991; Kliot, 1994). Well into the twenty-first century, Turkey will have invested some $30 billion to build dams, hydropower stations, and an irrigation canal system to bring more than 1.6 million hectares into cultivation. The flow of the Euphrates over the border into Syria may well be halved. Iraq and Syria have protested Turkey's unilateral action, and Iraq has claimed that it will suffer significant harm due to disruption in its established patterns of use.[6]

In 1987, Turkey, borrowing a leaf from Judson Harmon, guaranteed Syria a transborder flow of 500 cubic meters per second (cumsec), out of comity or good neighborliness. This was not an acknowledgment of a Syrian right. Syria, in 1990, in an agreement with Iraq, guaranteed Iraq 58 percent of what it received from Turkey (note that Turkey dealt in an absolute amount, and Syria, wisely, in a proportion). In response to the protests of the two downstream riparians, Turkey has suggested that a regional water market be established to allocate water according to economic studies of feasibility and productivity. Turkey must assume that its GAP projects would demonstrate higher returns than any use or project currently underway in Iraq and Syria (Hamidi, 1998).

There are many precedents in international watercourses for pricing the resource, directly or indirectly. In 1924 and 1925, in an exchange of notes between the Governor General of the Sudan (British) and the Governor of Eritrea (Italian), an allocation of the annual flow (virtually restricted to the flood season) of the Gash River was agreed upon. The government of the Sudan was obligated to pay annually to the government of Eritrea 20 percent of the value of agricultural production in the Gash Delta in the Sudan and £50,000 (Fisseha, 1981:190). In 1947, Egypt agreed to pay Uganda nearly £1 million in partial compensation for adding a meter to the crest of the Owen Falls Dam at Jinja in Uganda. Uganda did not need the additional storage capacity, the primary effect of which was to augment the flow of the White Nile during the dry season to Egypt's benefit but to lower power production at the dam site (Wilson, 1967:5). Israeli experts have suggested that Israel might be able to purchase water from Egypt in return for providing the technology to help Egypt reduce water use without sacrificing existing uses (recall that acquired rights militate against such innovations)(see Kally, 1994). In the same vein, the Metropolitan Water District of Southern California in 1987 paid for lining irrigation canals and other works in adjacent agricultural districts in exchange for the water saved (Rosegrant, 1995). Such direct payments and compensation will be the glue of regime maintenance.

INCENTIVE CONSISTENCY

It is to be hoped that the negotiating parties will not have consistent incentives; incentive inconsistency may be the functional equivalent of the veil of ignorance, yielding protection and recognition of others' rights in order to protect one's own. Because regimes nearly always contain conflicting norms, it is best that the parties to the regime have incentives to uphold norms that are in conflict. If a riparian's interests are served exclusively by one norm, then it can afford rigidly consistent postures any time its interests are at stake. Thus, what we might call a pure downstream state, one that only receives transboundary water but gives none away, can rigidly adhere to the principle of prior appropriation.

Egypt is such a state (not quite so pure are Mexico, Iraq, Uzbekistan, and Vietnam). Not only does all its water come from across its borders, but it has virtually no rainfall to compensate for any reduction in river flow. A pure "upstreamer" can afford to espouse the principle of absolute territorial sovereignty. Ethiopia is a pure upstreamer. By most counts (see Chapter 4), Ethiopia has a dozen international watercourses that flow *from* its territory, while none flow in (again, not quite so pure are Guinea and Turkey).

Within decades of having espoused the Harmon Doctrine, the United States was obliged to back away from it in negotiations with Canada over the Columbia River in which the United States is the downstream riparian. At the same time, Canada, having invoked prior appropriation on the Waterton and Belly Rivers, felt obliged to move away from that position with respect to the Columbia (McDougall, 1971:272). States in midcourse on a transboundary river or which share a river as a boundary will benefit from inconsistent incentives. The debate over legal principles has been and will continue to be dominated by those with consistent incentives, and it is those states that have forced the watering down of the ILC articles to nonoperational pap.[7]

HARMON IS DEAD; LONG LIVE HARMON

Among other things, regimes protect against unilateral and potentially harmful actions by specific riparians. The ILC Convention calls for "due diligence" in ascertaining beforehand if one's actions are likely to cause harm, and, in any event, to notify concerned parties what the actor intends to do. It has been shown, time and again, that when the national interest is at stake, or when flouting international norms appears costless, unilateralism prevails. Turkey's President Suleiman Demirel in 1992 opined that Turkey's water is like Saudi Arabia's oil, to be done with as Turkey sees fit and with no more international displeasure than was exhibited toward Saudi Arabia after 1973 (as cited in Waterbury, 1994:57). Turkey is by no means alone. In the middle 1950s, Israel began to construct its National Water Carrier to siphon water from the Jordan Basin at an off-take at Lake Tiberias and to move it far south to the Negev desert. Israel had no agreement with the other riparians (Jordan, Lebanon, and Syria). Moreover, this project involved an extrabasin transfer of water upon which international law looks askance. Syria, in turn, built scores of microdams in its portion of the watershed of the Yarmouk River, a major tributary of the Jordan, without any understanding with Jordan, nor with Israel with which it is at war. Similarly, India, in 1975, completed the Farrakka Barrage on the Ganges to divert dry-season water into the Hoogli River in order periodically to flush the port of Calcutta. The diversion deprived Bangladesh of what it had traditionally used. A five-year accord was negotiated in 1977 but lapsed after 1982 with India playing hegemon in the Ganges (see Salman, 1998; Crow and Singh, 2000). The People's Republic of China has proceeded unilaterally with hydropower development in the upper reaches of the Mekong without any warning to, let alone an accord with, the downstream riparians (Cambodia, Laos, Thailand, and Vietnam). One senses incipient upstream unilateralism brewing in Ethiopia (Ethiopian leaders closely follow Turkey and the GAP) and Kyrgyzstan.[8]

I noted in the previous chapter the seeming paradox that downstream states can be guilty of unilateralism as well (Huffaker et al., 2000:266). This is an important point and at the very heart of the concept of the community of interest. If use begets rights, then *new* use may beget *new* rights. Thus, if a downstream state undertakes new projects that substantially change the way in which it uses water, that has serious implications for upstream interests, especially if a regime provides for periodic reassessment in light of new needs and practices. Due diligence and formal notification must be exercised by downstreamers as well as upstreamers. I should stress that I am applying my own legal reasoning here. Egypt, for example, rejects, on the basis of its location in the watercourse, any potential to cause harm equivalent to that of the upstream states. Raj Krishna, assessing World Bank criteria, notes that there is inconsistency in Bank policy, which aspires, today, to treat all riparians, regardless of location in the watercourse, equally. Operational Directive 7.50 essentially exempts downstream riparians from the obligation of prior notification in the event of undertaking major works on its part of the river (Krishna, 1998:42).

In the late 1970s Ethiopia protested Egypt's plans to build a freshwater canal, out of basin, and into the Sinai peninsula. Ethiopia charged Egypt with both nonnotification and out-of-basin transfer. Construction of the canal, known as the Peace Canal, has gone ahead, and Ethiopia, despite a change in government, has continued to protest. Egypt defends its action on three counts: the water will not go to Israel; it is in-basin as satellite imagery has revealed an ancient and dried-up branch of the Nile that flowed through the Sinai, and the water to be moved is not really Nile water but recycled drainage water. This is precisely the kind of legalistic dancing that the Chayeses cite approvingly as indicating the general proclivity to adhere to international norms.

THE TIME NEEDED

It requires a long time to negotiate and conclude treaties governing international watercourses (Bingham et al., 1994:135). One of the arguments for data-gathering and exchange is that this builds communities of experts who can coax along dithering politicians. The 1959 Egypto-Sudanese agreement was somewhat exceptional in that it was concluded in a remarkably short period of time. It did, however, require a change of government in Khartoum (by peaceful coup d'état), and the advent of leadership eager for Egypt's support, to bring the process to a close. More common are processes spanning decades. Even seemingly simple dyadic negotiations can take a long time. The 1986 Lesotho-South African agreement came at the end of a very long process. It may not matter that the units involved enjoy high levels of development, or even that they may be able to appeal to a recognized higher authority.

The wrangling among the riparians of the Colorado Basin is famous (and includes Mexico), and Canada has faced similar problems. A dam built in 1967 on the Mackenzie River in British Columbia caused damage to the Peace Athabasca Delta downstream. Twenty years later, after the formation of multilateral investigation teams, a liaison committee, and, in 1981, the Mackenzie River Basin Committee, the parties—three provinces and two territories— had *begun* negotiations on seven bilateral agreements. A federal inquiry report of 1985 warned (as cited in Kennett, 1991:93): "These and other festering problems need to be resolved before some unilateral action in an interjurisdictional watershed precipitates *irreconcilable conflict*" (emphasis added). These are surprising words coming from Canadians talking about Canada.

Perhaps less surprising is conflict of a similar nature within India's federal system. From 1924 to 1974 there was a quasi-hegemonic solution imposed on one of India's trans-state watercourses, the Cauvery River, draining the states of Karnataka (upstream) and Tamil Nadu (downstream with an important delta). This fifty-year agreement, engineered while Great Britain was in effective control of India, was enforced to allocate water between the two states. When the agreement lapsed in 1974, the stage was set for unilateral action by Karnataka. Independent India, having anticipated this and other trans-state water disputes, had enacted in 1956 the Inter-State Water Disputes Act, which called for setting up water tribunals in disputed watercourses. Such a tribunal was created for the Cauvery. In 1991 it ordered Karantaka to guarantee annual releases of 205 billion cubic feet to Tamil Nadu. This is a gross reduction in the downstream state's acquired rights under the 1924–74 regime. It is estimated that rice cultivation in the delta requires about 500 billion cubic feet of water (see Aiyar, 1998; Damodaran, 1996). The water is to be released according to a weekly schedule, with deficits in any one week added to the releases in the following weeks. Karnataka has periodically violated both scheduled releases and the gross amounts to be released. The issue was eventually kicked up to India's Supreme Court, which returned it to the Cauvery Water Dispute Tribunal.

Transnational water disputes are even more intractable. Since the early 1970s there has been talk of the need for a Euphrates regime but little more. Attempts to bring about a basin-wide accord in the Jordan have sputtered along since the early 1950s. They came to an abrupt and near-disastrous halt in 1964 when the nearest thing to a "water war" broke out between Israel and Syria. Jordan and Israel maintained a secret dialogue over shared water issues after 1967, and Jordan and Syria in 1985 reached an as-yet unfulfilled agreement on the Yarmouk, which serves as their common border. After the signing of the Oslo Peace Agreements, Israel and Jordan negotiated a formal water-sharing accord, but neither Syria nor the Palestinians were party to it.

India and Pakistan took well over a decade to negotiate the Indus Basin Treaty, signed in 1960, effectively dividing the river and its tributaries between them rather than putting them under joint administration (there is no Indus Basin Authority). This was achieved only with the determined involvement of the World Bank and the direct participation of its president, Eugene Black. As noted above, there is no basin-wide accord among Nepal, India, and Bangladesh on the Ganges-Brahmaputra. India acted unilaterally to put in operation the Farrakka Barrage in 1975, then in 1977 the Janata Party government concluded the five-year accord with Bangladesh which the subsequent Congress Party government allowed to lapse. Another accord between India and Bangladesh was signed in 1996, but Nepal has consistently been left out. Even when formal, basin-wide accords are negotiated, they frequently remain ineffective and ignored, especially when they involve several riparians (Rangely et al., 1994).

A breakthrough may have nothing to do with new information or more technical analysis or even a change in the natural regime itself. The breakthrough between Egypt and the Sudan in 1958 and between India and Bangladesh in 1996 came about because of a coup d'état in Khartoum, bringing pro-Egyptian leadership to power, and a change of government in India that allowed Chief Minister of West Bengal, Jyoti Bassu, to broker a deal between New Delhi and Dacca.

REGIMES BEGIN AT HOME

Without any formal understandings among riparians on sharing, allocation, or quality of water, significant steps can be taken *within* each riparian state that may enhance future prospects for basin-wide cooperation. Moreover, these steps will, in and of themselves, be beneficial to each riparian. International regimes, as Cooper (1997) stresses, can be no more effective than the national policies and institutions that will have to induce change in domestic structure and behavior.

1. *Each riparian encourages every domestic agency or unit that extracts and/or supplies water to apply real or surrogate prices that at least cover the costs of extraction, delivery, operation, and maintenance.* This means that ministries of irrigation or public works, regional development agencies and authorities, water users associations, municipal water supply companies, and so on, must adopt such pricing policies. Given water's life-giving and sustaining functions, these agencies will have to be monitored by regulatory boards that represent among others, the major consumers.

Bulk or metered charges, estimated use charges, service fees, or indirect water taxes on drainage water or water-intensive crops may be used for cost recovery and system maintenance. This means that in the agricultural sector water will be treated as a factor of production, like land, labor, and capital, and that its scarcity value, relative to other factors, will be reflected in its price or service charges.

2. *Each riparian adopts (or maintains) no policies that would discourage technological innovation and the constant search for more efficient water use.* Just as there are sunset industries that appear in the cycles of technological innovation and transfer, so too are there obsolescent uses of water. All policies should aim at revealing those uses so that water can be absorbed into activities that use it more efficiently and with higher returns. "Switching" within sectors, especially agriculture, should occur. In a study of Senegal's agricultural sector in the face of a secular decline in rainfall and increasing desertification, Venema and Schiller (1995) devise a simulation model that yields a viable agricultural development strategy that is compatible with reduced amounts of water and sustainable. It will, if implemented, require considerable switching from current practices.

It is important to note that there is debate about how much change in water demand can be brought about through pricing. One "school," most forcefully represented by David Seckler, posits that water in most agricultural systems is used now about as efficiently as it can be. In this optic, policies aimed at water demand management will fail. What is needed are practical steps to enhance supply (capturing more surface or groundwater, or recycling more of what we use). Against this position are those who advocate demand management through pricing as an effective means to reduce and/or shift demand. This debate is nicely summarized in Bingham et al. (1994). The two policy principles under (1) and (2) alone would make any basin-wide negotiations easier, because each riparian could come to the table with a legitimate claim to having established "best practice" utilization of water. Such practice would be beneficial to the riparian economy whether or not any regime emerges.

3. *A second level of stage-setting for formal cooperation might consist in an independent monitoring and assessment unit, representing all riparians, to report on progress toward (1) and (2).* Such a unit could be staffed by a panel of neutral, third-party experts, contracted by the riparians to carry out the assessment. The results of the assessment would become a public document. There would be no mechanism to sanction footdraggers or to reward overachievers. The relative performance of the riparians would become public and part of the cumulative knowledge that might one day be used in negotiations.

4. *All riparians would seek to harmonize and standardize their laws and regulatory regimes affecting water use and environmental protection.* Both the process of harmonization and the actual application of the rules, laws, and regulations could be monitored by a joint unit constituted by the riparians themselves. The possibility of the citizens of riparians having access to the courts of any given riparian, to seek legal redress of violations of laws that are commonly applied, should be strengthened. If two riparians apply the same laws on industrial pollution, then it should be possible for citizens of one riparian suffering from pollution originating in another to sue for damages in the latter's courts.

Point (4) has direct implications for (1) and (2), because regulatory regimes have implications for the relative costs of water use and delivery. In this respect, each riparian should be subject to the same constraints. In the event of basin-wide negotiations, this will reduce fears that any given riparian will free-ride on the others.

CONCLUSION

Supranational regimes, no matter how needed, will not fall like rain from the heavens. The necessary, but far from sufficient, initial condition is that there be some consensus among the potential parties to a regime that it is to be preferred to the status quo. Not all parties need embrace that consensus with the same enthusiasm. Those who are most enthusiastic must devise the means to compensate the least enthusiastic for their participation. That may mean a given party will absorb most of the up-front costs or help the less eager in areas that do concern them. But even if the necessary condition is met, it is not likely that a lasting regime will emerge, and those that do will face the constant threat of theatric adherence or outright defection.

Regime creation is a long-term process. It will typically span the lifetimes of several governments, the path toward it will be littered with setbacks, and the basic national interests of some of the key players may change radically (think of Pakistan breaking into two nations in 1972 or Russia giving up control of the five Central Asian republics). Given that kind of perspective, preparation for regime building should begin at home. Parties truly committed to supranational cooperation should begin to build domestic institutions and economic practices that will facilitate the process of negotiating a regime and monitoring its implementation. An extreme example is the European Union, which over the decades of its formation has induced uniformity in political and economic institutions and policies among its member states. Fortunately, in river basins that degree of uniformity is not required, but consistency in policies affecting the economic value of surface water and in environmental protection are. Comparing progress toward common policy goals in a nonbinding, loosely structured way can constitute the first step toward more formal supranational institutions.

As more formal steps are taken, there may be an intermediate phase of negotiated accords. The Chayeses (1995:226) cite the virtues of "framework protocols." The Barcelona protocol on Mediterranean pollution of 1976, or the Montreal protocol on cfcs of 1989, are both flexible and nonbinding, and serve as frameworks for consensus building. The ilc Convention of 1997 is a framework agreement. Unfortunately, although it has been adopted by the UN General Assembly, few of its articles received universal support, and on many points abstentions were the norm. The process of formal adherence and ratification does not look promising.

Within the basin, subsets of riparians may be able to reach accords because their shared interests are more obvious and the disparities in benefits resulting from cooperation less. As Lichbach stresses: "The subsetting of a large community into a series of small communities is therefore another possible solution to the cooperator's dilemma. The optimal CA [collective action] organization is large in size but decentralized in structure" (1996:187). It is important to note that Lichbach thinks this is so because *social constraints* are more likely to come into play in smaller groups than in larger ones (1996:199). The latter he sees as being bound by contract, not by norms. This notion does not work in the Nile basin. I argue that smaller subsets of riparians are more likely to reach *contractual* understandings, with less asymmetrical costs and benefits, than the basin as a whole. Just as there is no community of Nile states, there are no communities of subgroups. Interest-based contracts are simply more likely to be negotiated at the sub-basin level than at the basin-wide level.

I have suggested that economics provides "clean" criteria by which to decide on alternative uses and allocations of water. The simple fact is, however, that politics is often trump and never absent from regime creation in transboundary watercourses. Bates (1997) saw politics, both domestic and Cold War, as essential to an understanding of the rise and fall of the International Coffee Organization. In his comparative work, Le Marquand (1977) found politics paramount in most water negotiations. Journalistic accounts of the Cauvery dispute in India see it as resulting essentially from the ambitions of politicians at the state level. Everyone may agree on the economic logic of cooperation but still not take the first steps toward regime creation. At the same time, the kind of politics that Robert Putnam (1988) sees in two-level games may provide a useful way of selling regimes to reluctant domestic constituencies, and of winning concessions in the rules and regulations of a regime by stressing the harm that may befall key domestic supporters.

In regimes based on voluntary compliance, there are no criteria of allocation of resources and of costs and benefits that all participants will accept as just and fair. There may be some that all will accept as relevant. Debates over equity, like the search for perfect knowledge of the resource, will lead to no definitive conclusions, and may well be used as delaying devices by those unconvinced that the status quo needs to change. An initial allocation is likely to be somewhat arbitrary; it will be more like selling a rug in a marketplace than carefully delineating shares according to objective criteria. Equity will be served to the extent that all parties to the regime must come away with something they did not have before. Some may come away with a lot, and some with very little, but that may be sufficient to move ahead. It is important that the initial allocation of water resources, the assignation of property rights, not be graven in stone. The possibility of trading rights must be open, even though in reality trading will seldom be more than a tiny proportion of the total quantities involved. The ability to trade softens the image of a once-and-for-all distribution of a scarce good.

As the rules and regulations of a regime are given shape, the designers would do well to build in provisions for periodic review and renegotiation in light of the changing interests of the parties and of the changing nature of the natural regime governing the watercourse. One wants to avoid locking in participants forever; if they think that future alterations are not allowed, they may either fail to join or simply leave when they feel their interests are being hurt. An incentive must be given to negotiate adjustments.

It is the case today, at the beginning of the twenty-first century, that many of the elements we have identified as necessary for collective action and regime formation are in place in the Nile basin. There is a collective recognition of the need for cooperation; there is an older RBO linking Egypt and the Sudan and new RBOs provided essentially by concerned donors, centered in the World Bank; there is a growing technocracy that may have a vested interest in cooperative action; and there are limited steps to build domestic legislation and institutions that would reinforce basin-wide cooperation. Despite that, the status quo prevailing since 1959 is not yet under threat. The continued combination of a quasi-hegemon devoted to the status quo, the use of riparian resources for other priorities (war prominent among them), and the residual indifference of a number of riparians to the benefits of cooperation have rendered a new regime at best a dim and hazy prospect.

Chapter 3 The Three-Level Game in the Nile Basin

In the past century and a half, corresponding to the period in which control over the supply of the Nile's water became a concern of the lower riparians, there have been three distinct regimes in the basin, and we may be entering a fourth, that is, one of basin-wide, voluntary cooperation. Throughout this period the power games in the basin

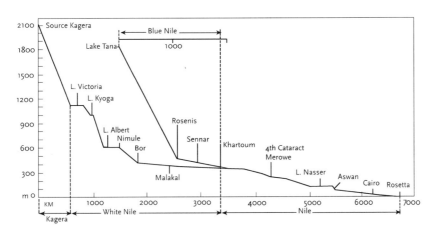

Figure 3.1. Slopes of the White, Blue, and Main Niles
Source: Giuliano Ferrieri et al., *Il Nilo* (Novara, 1978), p. 14.

have been played at three levels. First-level players have been great, extrabasin powers: Great Britain, France, and Italy, followed by the United States and the Soviet Union. Second-level players have been the national entities of the basin. Since the end of the Cold War they have become the primary players in the power games. The third-level players have been various kinds of subnational interests and groups. They were prominent in the first half of the period under examination, but with two exceptions they have been eclipsed roughly since the 1930s.

Throughout most of the nineteenth century, rival colonial powers jockeyed for position in the basin. As noted, the principal actors were Great Britain, France, Italy, and, until 1882 when Britain asserted direct control over it, Egypt. These powers interacted with and manipulated substate actors defined by region, religion, and blood. The only entity enjoying "statehood" was Ethiopia, but it was weak and internally fragmented, offering ample opportunity for outside meddling in its affairs. This was a regime without a hegemon in which the rival forces succeeded mainly in preventing any other force from tampering with the natural flow of the Nile.

The second regime prevailed for the first half of the twentieth century, when quasi-hegemonic authority in most of the basin was established under the auspices of Great Britain. The local actors now were nation-states under colonial protection. Even Ethiopia briefly lost its sovereignty to the Italians during this period. A regime favoring Egyptian uses of the Nile was established.

The third regime was the result of the Cold War and covers the period 1945–89. By the early 1960s the British presence was negligible in the basin, and U.S.-Soviet rivalry, played out through shifting client states in the basin, led to a regime that in fact reinforced the existing arrangements in favor of Egyptian, and, to a greater extent after 1959, Sudanese interests.

Under the latter two regimes only two significant subnational forces were players in the game of regime maintenance: the peoples of the Southern Sudan and the Eritreans.

Despite Italy's lengthy control of Eritrea and its short occupation of Ethiopia, the latter was never brought into any stable regime. Its ability to resist any hegemonic solution was a source of frustration and anxiety to the Egyptians and to the British. The thwarted attempts of the British to form a regime that included Ethiopia is a major focus of this chapter.

In British imperial strategy, India and the sea-lanes to the Far East were the fundamental assets of empire, to be carefully defended against encroachment by France, Prussia/Germany, or Russia. The Suez Canal linked physically and strategically the Indian subcontinent to the Nile basin. By the last quarter of the nineteenth century Great Britain not only owned a controlling interest in the Compagnie Universelle de Suez but also had direct control of Egypt itself after 1882. Egypt and the canal, in British strategic thinking, constituted vulnerable links in the sea passage to India and thence to China. Britain adopted a strategy of defense in depth which entailed gaining control over the eastern flank of the canal, meaning Palestine, and eventually all that lay south and west of Egypt: Libya, Sudan, and the equatorial lakes. The missing piece in the puzzle was Ethiopia, more specifically Lake Tana in the western Ethiopian highlands.

The British conceptualization of defense in depth went beyond controlling real estate to enhancing its productivity. First and foremost this meant building and protecting Egypt's irrigated agricultural sector and augmenting the production of cotton. After World War I, the Sudan entered into the same equation. At the turn of the century, Sir William Garstin, a senior engineer in Egypt's Department of Public Works, carried out a survey of the upper Nile basin. The motivation for this study was British concern over ensuring Egypt's agricultural water supply. At that time the water supply of the White Nile, flowing out of the equatorial lakes, was unchallenged. British suzerainty was complete and territorially uninterrupted from Entebbe, Uganda, in the south to Alexandria, Egypt, in the north. The problem was and is that relatively little of the total flow of the Nile comes out of this source. By contrast 85 percent on average comes out of the Ethiopian highlands, over which Britain had no direct control. In many ways Ethiopia was to the Nile basin in those years what Afghanistan was to the Indian subcontinent. Ethiopia was fought over but, except for one brief interlude under Mussolini, never conquered. The French, British, and Italians all maneuvered for position in Ethiopia. In the final analysis the British were glad to trade influence in other parts of Ethiopia for supremacy in the Lake Tana basin.

The reasons were simple. From the time of Sir William Garstin's pioneering work at the turn of the century, until after World War II (when it no longer made any difference to the British), Lake Tana was seen as the natural controlling reservoir for the Blue Nile, the spigot, as it were, for 85 percent of Egypt's water.[1] Moreover, because Egypt's cotton crop demanded most water over the summer months, when the Nile in Egypt is normally low, Lake Tana, as a reservoir, could be used to release "timely" water. On an even grander scale, Tana was seen as crucial to the so-called Century Water Scheme, directly derived from Garstin's survey, that identified a number of storage projects in the Upper Nile

Basin that would hold sufficient water to guard against a succession of low floods likely to occur only once in a century. With that three-part calculus firmly in mind, and well understood by Britain's rivals, the game was on.

THE TANA CONCESSION

The realist school of international relations has as its premise the proposition that the international system, which embodies a distribution of resources and power, determines in fundamental ways how any individual state will act in its own interests in the prevailing international regime. It would be hard to argue that any state, let alone domestic interests *within* any state, could act in ignorance of that reality. There are variants of realist understandings of international relations. The terms in which Gideon Rose (1998:149; see also Legro and Moravcsik, 1998) has summarized the tenor of what he calls the "offensive realist" school are broadly applicable to all actors in the Nile basin throughout the past century and a half:

> Domestic differences between countries are considered to be relatively unimportant, because pressures from the international system are assumed to be strong and straightforward enough to make similarly situated states behave alike, regardless of their internal characteristics. According to this view, foreign policy activity is the record of nervous states jockeying for position within the framework of a given power configuration. To understand why a state is behaving in a particular way, offensive realists suggest, one should examine its relative capabilities and its external environment, because those factors will be translated relatively smoothly into foreign policy and shape how the state chooses to advance its interests.

Delete "relatively smoothly" and this paragraph applies well to all the actors in the Nile basin, whether states, tribes, or religious sects.

In the nineteenth century, before anything resembling a state system had emerged in Africa (Herbst, 2000), domestic constituencies played a prominent role in all issues affecting the Nile. These were not "conventional" constituencies, like coffee growers in Colombia; rather, they were religious, racial, and tribal constituencies. European powers, seeking strategic advantage, formed tactical alliances with selected constituencies and entered indirectly into local political contests. This was how the game was played in Ethiopia, or Abyssinia, as it was then known.

The national (second-) level players in the game were the Khedives (Ottoman governors) of Egypt, the Mahdi in the Sudan, and the emperors in

Ethiopia. The first-level players were Great Britain, France, and Italy. The third level was crowded with Eritreans, Tigreans, Somalis and Afars, Oromo and Amhara, Beja, Nubians, Muslims and Christians, local potentates, and quasi-autonomous governors. No one had firm control of anything. Allies were bought, sold, and betrayed with disconcerting frequency.

The stakes for the level-one players were control of real estate for its own sake. This was the quest of the Italians, described wonderfully by Bahru Zewde (1994:56) as combining "the vigour of youth with the desperation of the late comer." For the British it was Lake Tana and the diversion of coffee and gold trade from southwest Ethiopia into the Sudan, which, after 1896, was under British rule. For the French it was the railroad running from Djibouti to Addis Ababa, and, they hoped, all the way west to the Nile and the Sudan. France wanted to control all the trade coming out of eastern Ethiopia, and perhaps one day to nibble into British-dominated trade in the southwest.

It was, of course, an option for European powers to seize what they wanted, but that was not an altogether rational strategy. The British did it in the Sudan, in 1896, when General Kitchener defeated the Mahdist forces and avenged the death of General "Chinese" Gordon. But what had the British seized? Nothing very much as there was no unified state administration of the Sudan. When Great Britain moved militarily into Egypt in 1882, it took control of a sophisticated administrative system that counted people, put them in uniform, made them pay taxes, and conscripted their labor. In the Sudan, the British seized Khartoum and Omdurman and then, at considerable cost, had to build a state system and, in fact, an economy. By the time the British focused on Lake Tana they were no longer interested in seizure.

So too the French, although their conquest of Morocco still lay ahead. Fashoda (see the Introduction) signaled the costs of military conquest. The crushing defeat of the Italians at Adwa in March 1896, at the hands of the Ethiopians, was graphic evidence of the perils of fighting in the Abyssinian highlands. (The message the British took from Adwa was that they had better seize control of the Sudan before the Ethiopians did, but, by the same token, avoid the highlands at all costs.) The Italians seemingly learned nothing from Adwa except to apply more force. They waited nearly forty years and then invaded Ethiopia. Their seizure of the highlands lasted about five years. The desperation of the latecomer, indeed.

Realist lessons were learned in those years that have rung remarkably true in recent decades. Khedive Ismail of Egypt, the Ottoman governor who presided over the completion of the Suez Canal, the expansion of Cairo, and the first performance of *Aida,* had ambitions to extend Egyptian influence throughout the Sudan and all of northeast Africa. These ambitions set him and Egypt at direct odds with the Abyssinian and Christian Emperor Johannes, who appealed to European Christian powers to protect him against Muslim expansionism. As Zewde (1994:51) writes: "The European powers did not find the theme of Christian solidarity very convincing. To them, Muslim though it was, Egypt offered more opportunities for trade and investment than Ethiopia did, for in economic terms Ethiopia was a relatively unknown quantity."

Pursuit of material interest, as the realist school would predict, predominated in the games played in the Horn of Africa. The Suez Canal raised Egypt's strategic saliency to such an extent that Britain, faced with a protonationalist movement in the country, felt obliged to occupy Egypt militarily in 1882. Three quarters of a century later, the United States and the USSR played cold war games in the Horn. There was never any doubt that Egypt was still the prize, while Ethiopian territory and markets, let alone friendship, were always expendable.

The defeat of the Italians at Adwa was a shock to European powers, one compounded by the success of the Mahdists in the Sudan. Great Britain was willing to help Italy recoup its losses and, in 1900, to take control of Eritrea, where it remained ensconced until 1941. Nearly simultaneously, the French began construction, in 1897, of the railroad that would link Djibouti to Addis Ababa by way of Dire Dawa. The railroad was completed in 1917. The various spheres of influence were delimited in the Tri-partite Treaty of December 1906.

Once in control of the Sudan, the British wanted to delimit its frontier with Ethiopia and to lobby for a concession on Lake Tana. In 1902 a Treaty was signed between Great Britain (King Edward VII) and Ethiopia (Emperor Menelek II) that differs significantly in its English and Amharic versions. The controversial clause in its English version reads (Ministry of Foreign Affairs [Ethiopia], 1993): "His Majesty the Emperor Menelek II, King of Kings of Ethiopia, engages himself towards the government of His Britannic Majesty not to construct or allow to be constructed any work across the Blue Nile, Lake Tsana or the Sobat, which would arrest the flow of their waters into the Nile except in agreement with His Britannic Majesty's government and the government of the Sudan."

This clause has since been sporadically invoked by both Egypt and the Sudan to warn Ethiopia that it is bound by treaty not to impede the flow of the Nile. Ethiopia has argued that the Amharic version clearly binds Ethiopia to seek agreement from Britain *as the effective suzerain of the Sudan.* It does not oblige Ethiopia to seek agreement with the sovereign nations of the Sudan and Egypt. It should be noted parenthetically that neither the English nor the

Amharic versions mention the Teccaze River, which flows into the Atbara in the Sudan and which is not part of the Blue Nile watershed.[2]

Emperor Hailie Selassie reflected on Ethiopia's obligations arising from the 1902 Treaty and insisted that they bore only on the authorities of the British Sudan, an entity that came to an end in 1956. The emperor noted that by the terms of the 1902 text, Great Britain was to pay Ethiopia 10,000 guineas annually, but that from 1902 to 1931 no payments were made. When, as emperor, Hailie Selassie sought to collect back payment, he was told that until and unless Great Britain received the concession on Lake Tana no arrears would be forthcoming (Hailie Selassie, 1976:139, 145).

The 1902 Treaty provided for a trading station for the British Sudan at Gambella in the Ethiopian salient jutting into the swamps of southern Sudan. It also granted Great Britain the right to construct a railroad, presumably from Khartoum and Wad Medani in the Sudan, across Ethiopian territory to link the Sudan to Uganda. That railroad never came into existence, but a railroad from Khartoum to Port Sudan on the Red Sea was built. In 1910 it was extended to Wad Medani in what was to be the heart of the Gezira cotton irrigation scheme. With its construction, cotton could be transported by rail from the Gezira scheme to Port Sudan for shipment to the mills of Manchester and elsewhere. Lake Tana came then to be seen as the key to providing water for cotton cultivation in the summer months, March to July, when the Blue Nile is at its lowest flow (Zewde, 1976:40; McCann, 1981).

In the early 1920s, the regent Ras Tefari, later to become Emperor Hailie Selassie, was maneuvering for the supreme position of *negus*. He had strong rivals in the governors of the two Blue Nile provinces of Gojjam and Begemder. Ras Tefari saw the issue of a British concession on Lake Tana as a potential Trojan horse, allowing western penetration into these traditional seats of Amharic power, thereby giving ammunition to his rivals. He saw the issue of the concession as the ticket to British backing for his ascent to the throne; he feared that Ras Hailu Tekle Haymonot, governor of Gojjam, might try the same ploy to further his chances. Ras Tefari presented himself to British emissaries as a would-be reformer, locked in a struggle with reactionary forces. Were he to become emperor, and were he to bring to heel the two powerful governors, then he could grant the British the Tana concession. Negotiations proceeded during 1922 and 1923. Lord Curzon offered annual payments to the regent and an annual rental once the storage project was completed. But the British sent mixed signals, opposing for a time Ethiopian membership in the League of Nations and continuing negotiations with Ras Hailu. Talks with Ras Tefari were broken off at the end of 1923.[3]

In 1926 Great Britain completed construction of the Sennar Dam, upstream of Khartoum on the Blue Nile. With that, part of the problem of regulated water supply to the Gezira scheme was solved (although Sennar had only seasonal storage capacity and could not protect the Gezira against a very low flood, let alone a succession of low floods). Sennar deprived the Tana quest of some of its urgency. Great Britain began increasingly to wager on Italian influence in Ethiopia to secure its interests.

In an exchange of notes in November 1924 and June 1925 between the British governor general of the Sudan and the Italian governor of Eritrea, an accord was reached on the flow of the Gash River into the Sudan from Eritrean territory. The Sudan would pay Eritrea an annual rent plus a share of the value of agricultural produce in exchange for a guaranteed flow. More significant for this chapter is the fact that Great Britain conceded to Italy an economic sphere of influence in all of western Ethiopia in exchange for Italian good offices in promoting an Ethiopian concession to Great Britain to develop Lake Tana as a storage site (Zewde, 1976:151–52; Fisseha, 1981:190). By this time, Ethiopia had been admitted to the League of Nations, and the British-Italian deal was seen as a gross infringement of Ethiopian sovereignty. It was at least clumsy, and, as an attempt to pressure Ethiopia into granting the concession, it backfired.

The regent sent a trusted emissary to the United States, who made contact with the construction and contracting firm of J. G. White. The emissary announced a privately financed project for $20 million to build a regulating dam at the outlet of the Blue Nile (Abbay) from Lake Tana. The message to Great Britain was clear: we will carry out the project with our own resources, and it will be under the sovereign control of Ethiopia. In 1928, Ras Tefari was crowned *negus,* and in 1929 he awarded the Tana concession to J. G. White.

Again, presaging a phenomenon that would reemerge in the 1970s and 1980s, the onset of the world depression in 1929 and 1930 significantly recast the objectives of Great Britain, and, to the extent that the depression facilitated Mussolini's coming to power in Italy, Italian objectives as well. It also revealed Ethiopia's Achilles heel, a backward economy with no international credit-worthiness. As the crisis deepened, Ethiopia sought commercial loans to finance the Tana project, using future revenues from power generation as collateral.[4] It did not work. Ethiopia's timing and credit-worthiness were equally bad. Ethiopia's bluff had been called, but it was a bluff that the same leader, in dramatically altered circumstances, would try after 1958 as Egypt and the Sudan agreed to the construction of the Aswan High Dam.

The British were no longer in a hurry for a concession. World prices for cotton were tumbling. Pragmatists in Whitehall proposed putting the Tana project on hold until cotton prices recovered and Egyptian and Sudanese water demand increased significantly. As the emperor sought to rescue the project, Sudanese officials called for more studies and warned that the Sudanese economy in its current state could not be squeezed for finances for the Tana project.

By 1935, the *negus* was concerned by the increasingly bellicose policies of Mussolini. He made one last attempt to hold a conference to award the Tana concession, and, assuming that both U.S. and British interests would be involved, to buy some support against Mussolini. Anglo-Egyptian officials declined to attend the conference, and J. G. White effectively pulled out of the project. McCann (1981) argues that with the collapse of this conference, Italy received a green light to invade Ethiopia.

The game, as I have noted, involved competition between Britain and France for shares of Ethiopia's trade. Once Britain had conceded to Italy an economic sphere in the western portions of Ethiopia, only the Gambella salient and trading post remained of interest to the British. Their hope had been to use barge transport through the southern swamps of the Sudan to move Ethiopian coffee, ivory, hardwoods, hides, and gold to the railhead at Khartoum and thence on to Port Sudan. The main alternative, and one increasingly favored by Ras Tefari, was the French-built railroad from Addis Ababa to the Red Sea at Djibouti.

As it turned out, the Baro-Sobat rivers and the Machar swamps were navigable only during the rainy season when the Gambella hinterlands and the southwest plateau were impenetrable. Although considerable amounts of coffee were exported through the Sudan, by 1935, 75 percent of all Ethiopian trade went through Djibouti (Zewde, 1976).

THE COMING AND GOING OF THE COLD WAR

After We rested for a little while, We went for a walk to the Nile with Chapman-Andrews. When We reached . . . [the] river that emerges from Our country, We were moved by deep feelings of nostalgia. In fact, We cupped Our hands, scooped up some water, and sipped a little.
—*Hailie Selassie* (1994:101)

In June 1940, after an exile in England following the Italian conquest of his country, Emperor Hailie Selassie traveled to Egypt and thence to Wadi Halfa in the Sudan on his way back to Ethiopia to lead the resistance to the Italians. It was at Wadi Halfa that he drank from the main Nile. By 1942, Mussolini's grip on Ethiopia had crumbled, and for two years Britain enjoyed near-direct control over the country. It was for the British too much too late. The dynamics of decolonization had already begun to make themselves felt. Although it was in a position to dictate the terms of a Tana concession, Great Britain no longer had the will or the capacity to defend the empire "in depth." No concession was ever awarded, although Ethiopia did build a regulating dam and small power station at the outlet of the Abbay from Tana in the early 1960s (see Chapter 5).

The game of the late nineteenth and early twentieth centuries was replaced by the Cold War, and, as Britain, France, and above all Italy, played diminished roles in the region, the United States and the USSR entered the scene in a major way. The Americans wanted no vacuum to develop as France and Britain granted their Middle Eastern and African possessions independence. The Truman Doctrine, announced in May 1947, showed U.S. determination to limit Soviet influence in Turkey, Iran, and Greece, and it was followed by the attempts under the Eisenhower administration to bring Arab countries such as Iraq, Syria, Jordan, and Egypt into a military alliance associated with NATO. That effort brought the Cold War to the banks of the Nile.

In July 1952, the Egyptian monarchy was overthrown in a military coup d'état led by General Mohammed Naguib and Colonel Gamal Abd al-Nasser. The latter in particular combined a strong suspicion of the former colonial powers and their ally, the United States, with a populist-cum-socialist outlook that, superficially, drew inspiration from the transformations that took place in the USSR. Despite the blandishments of John Foster Dulles, Egypt was not to be coaxed into a western, anti-Soviet alliance. Indeed, so frustrated was Egypt in seeking arms from the West, that, in 1955, it turned to Czechoslovakia as a conduit for Soviet arms. The Czech arms deal allowed the USSR to leapfrog over the anti-Soviet northern tier alliance of Turkey, Iran, and Iraq.

It also prompted the United States and the World Bank to withdraw from funding the Aswan High Dam. So long as Egypt was a prospect for membership in a western alliance, the United States and Britain encouraged the World Bank to explore ways to help Egypt construct this project to provide over-year storage of the Nile flood.[5]

The Czech arms deal transformed the western calculus and led to the cancellation of funding. Nasser retaliated in July 1956 by nationalizing the Suez Canal company in order to capture its revenues to construct the Aswan High Dam. The nationalization in turn prompted Great Britain, France, and Israel to collude in a military attack on Egypt in November 1956. Not only did Nasser emerge from this confrontation as the paramount leader of the Arab world, but it also thrust Egypt and other Arab countries squarely into the Soviet camp. The USSR took over funding of the dam, mobilized significant assistance for Egyptian industrialization, and began to play a major role in building the Egyptian armed forces.

Since Sudanese independence in 1956, Egypt and the Sudan had engaged in negotiations over the construction of the Aswan High Dam. The reservoir upstream of the dam would back well into Sudanese territory and displace populations living along the banks of the river. The talks were begun because the World Bank insisted on an accord before it would advance financing. For their part, the Sudanese were contemplating a second dam, after Sennar, on the Blue Nile. Egyptian consent would have to be attained before construction could go forward. The talks continued even after the World Bank withdrew its support of the Aswan High Dam. I will discuss the details of these negotiations in the next section. At this point I want only to note that the Sudan sought a comprehensive accord on the Nile with Egypt, one that would supersede the 1929 agreement, signed when both countries were under British control. For two years the talks progressed little, but in 1958, General Abboud led a putsch in Khartoum that brought to power a team that greatly admired Gamal Abd al-Nasser or at least sensed that the regional winds were blowing in his favor. With Abboud in power, the 1959 agreement was concluded. To the west it appeared that Egypt had brought the Sudan within the Soviet ambit.

Soviet influence in Egypt and the Sudan in the late 1950s is impressive only with hindsight. Nikita Khrushchev, the Soviet premier, was not an experienced statesman at this point, and developing influence in Egypt was in many ways the first Soviet experiment in a third world country. Nonetheless, the two second-level players at this point, Egypt and the Sudan, bargained under a Soviet umbrella. The USSR had no obvious stake in any particular outcome but presumably would back anything Egypt was willing to live with. Much the same situation developed between Syria and Iraq when, in the 1970s, they negotiated over the Euphrates River at a time when both countries were Soviet client states. In this arena, the USSR probably tilted slightly toward Syria where a major dam on the Euphrates was being constructed.

The United States had assets other than Egypt in the Nile basin. Foremost among them was Ethiopia, led since the 1940s by the pro-western Emperor Hailie Selassie. In the early 1950s the United States built a monitoring station and air base at Kagnew, outside Asmara, the capital of Eritrea. From this vantage, the United States, and NATO, could monitor all air and sea traffic in the

Red Sea and the Suez Canal. This asset was of such strategic value that the United States acquiesced in the absorption of Eritrea into Ethiopia in November 1962. The dominant logic of the Cold War in this and many other instances led to the suppression of movements for self-determination in former colonies and possessions around the world.

In fact the only domestic actors in the two-level Cold War game were ethnic groups in the southern Sudan and in Eritrea seeking autonomy if not independence. One year before independence, in 1955, black southern troops in the Sudanese armed forces mutinied in Torit and other garrisons in the south, setting off an insurrection and virtual civil war that has yet to end. Likewise in Eritrea, after 1961, two liberation fronts were launched to throw off Ethiopian suzerainty. It was thirty years before their objective was reached. Both the Soviet Union and the United States avoided directly aiding either of these movements because the regimes against which they were in revolt were potentially more valuable assets than were the insurgents in the Cold War chess game.

Like the countries of the basin themselves, the great powers saw dissident movements as convenient instruments for short-term destabilization of hostile governments, but they never figured as instruments in longer term strategic objectives. The risk of aiding minority movements was that of permanently alienating the majority even when the leadership of the majority changed. Moreover, aiding dissident movements that might lead to the breakup of newly sovereign, fragile states risked alienating many more developing countries than those immediately affected.

Aside from the insurrections and the resulting civil strife, there were no domestic actors, outside the armed forces, that played an active role in basin-wide politics. There was scarcely any trade among the most important riparians nor was there much investment. For all its historic claims to the Sudan (claims muted if not totally silenced under the Nasserist regime), Egypt invested very little in that country. Neither Egypt nor the Sudan had significant economic interests in Ethiopia. As a result there were no interest groups in any of the Nile countries that had positions to protect or promote in the markets and territories of the others. For this reason, and as noted in Chapter 1, the officials of the respective states could and did proceed as unitary actors with uniform strategic and economic objectives.

In many respects, the most auspicious moment for Ethiopia to assert itself in the affairs of the Nile came in 1960–61. It had successfully asserted its claims to Eritrea. It was the seat of the Organization of African Unity. It had strong backing from the United States, which sought a counterbalance to the growing

Soviet presence in Egypt. Ethiopia had rejected the 1959 agreement between Egypt and the Sudan as nonbinding on any of the other riparians. In 1958, Ethiopia invited the Bureau of Reclamation of the U.S. Department of the Interior to carry out an extensive survey of the Blue Nile (Abbay) and Teccaze watersheds with a view toward developing Ethiopia's water resources. The study was completed in 1964, at a time when Egypt and the United States (under Lyndon Johnson) were in a particularly confrontational mood over Egyptian and U.S. maneuvers in the Congo.

The Bureau of Reclamation study, in several volumes, identified over twenty major water development projects for irrigation and hydropower development. The gross abstraction of Nile water that these projects would cause, were they to be implemented, was estimated at over 4 bcm, or about 5 percent of the mean discharge of the Nile as measured at Aswan.[6] Just as Hailie Selassie had tried to outflank the British in the 1930s by going directly to a prominent U.S. firm with the Tana project, so too in 1958–64 did Ethiopia, with the active help of the United States, try to outflank Egypt and the Soviet Union by intimating that it might proceed unilaterally to develop its western watershed, thereby undermining the very logic of the Aswan High Dam. The message to Egypt was "you have the Russians and the Aswan High Dam, but we have the United States and, more important, the water."

In the final analysis, however, Ethiopia did not assert itself. It never followed up in the international community on its objections to the 1959 agreement, and, because of its weak economic and financial base, it could not follow up on the Bureau of Reclamation report. In the one project, at Finchaa (see Chapter 5), that resulted from the study, USAID made at best a modest contribution to construction. Clearly the study was intended to be part of Cold War theatrics rather than a blueprint for Blue Nile development.[7]

Cold War alliances forged after 1956 lasted until the early 1970s. Egypt remained firmly aligned with the Soviet Union while Ethiopia remained firmly aligned with the United States. All that changed dramatically and fairly swiftly after the death of Gamal Abd al-Nasser in 1970 and the overthrow of Hailie Selassie in 1974. Anwar Sadat, who succeeded Nasser as head of state, began to open the Egyptian economy to western investment and to reduce Egypt's reliance on trade with the USSR and Eastern Europe. He entered into hostilities with Israel in October 1973 in part to achieve a settlement that could end the state of war that had prevailed since 1948. A formal settlement was not produced until the Camp David Accords of 1978–79, but in the intervening years Egypt openly shifted camps in the Cold War.

So too did Ethiopia. In a putsch in February 1974, Colonel Menguistu Hailie Meriam overthrew Hailie Selassie, installed the junta, or *dergue,* and eventually proclaimed Ethiopia a radical, socialist republic. Soon Russians and Cubans were advising and training Ethiopians in the overhaul of the armed forces, agrarian reform, and state-led industrialization. The acquisition of this new asset compensated the USSR in part for the "loss" of Egypt. It also led the Soviet Union to abandon its erstwhile ally in the Horn, Siyyad Barre of Somalia. Soviet and Cuban help was crucial in thwarting the Somali thrust into eastern Ethiopia (from Djijiga to Harrar) in the Ogaden war of 1977–78. The United States, predictably, adopted Somalia as its influence in Ethiopia came to an end. By 1979 there had been a total reversal of Cold War alignments in the Nile basin. Curiously, the linkage with the original home of the Great Game remained strong: Afghanistan came under direct Soviet control in 1979. Ethiopia was thus the southern anchor of an arc (so dubbed by Zbigniew Brzezinski in the Carter administration) of radical states running through Aden, Iraq, and Iran (the Shah fell to the Islamic revolution in February 1979) to Afghanistan.

The U.S. alignment consisted of Egypt, Saudi Arabia, Jordan, and Somalia. Israel was a conspicuous but unacknowledged piece of this alignment. The Sudan, as is generally its practice, did not stray far from Egypt. Ja'afar al-Nimeiri had come to power in 1969 through a military coup and announced his regime's devotion to the principles of Nasserism. With the death of that role model, al-Nimeiry led the Sudan, as much as he followed Sadat's Egypt, away from the USSR and toward the United States. In an acid test, the Sudan supported Egypt's signing of the Camp David Accords with Israel in 1979 and the establishment of peaceful relations between the latter two countries.

With respect to Nile water issues, the new alignments had major consequences. The Ethiopian delegation to the UN Water Conference at Mar del Plata in Argentina in the spring of 1977 presented a country paper that was based largely on the Bureau of Reclamation survey of the western watershed. It warned that development of that watershed was an imperative for the economic health of Ethiopia. It called for good neighborliness in the basin but reserved its right to proceed unilaterally with water development projects (see Ethiopia, 1977).

In 1979, Egypt announced its intentions to construct a fresh water canal passing under the Suez Canal and out into the northern Sinai peninsula. It was to be called the "Peace Canal." Although supply of water to Israel was never mentioned, there were strong suppositions that Israel was the ultimate destination of the canal. Ethiopia denounced the project as a unilateral act and an extrabasin transfer of water from the Nile. The Ethiopian protest came at a time when rumors circulated that Ethiopia, with Israeli help, was planning to build dams on the Abbay.[8] On May 30, 1979, President Sadat threatened to go to war, implicitly against Ethiopia, to protect Egypt's water supply, and he proclaimed that in this respect defense of the Sudan was the same as the defense of Egypt.

Egypt and Ethiopia have consistently manifested radically different tactics in defending their Nile interests, and Egypt has indisputably been the more effective. Ethiopia has punctuated what might be called "aggressive silence" on Nile issues with short bursts of protest or denunciation followed by little or no action. It has somehow believed, both under the emperor and under the *dergue,* that the patent justice of its position would speak for itself. In all basin-wide fora, such as the Hydromet survey of the equatorial Lakes, the Undugu group of Nile states, and the subsequent Tecconile/Nile 2002 organization, Ethiopia generally preferred to participate as an observer rather than as a member.[9] Its political turmoil has meant that it has not built continuity among the experts who understand Nile issues, and, even more important, it has not "infiltrated" the international funding organizations that could help Ethiopia finance its plans. Its water technocracy has been affected over the years by political purges, politically driven migration, and, since 1993, by campaigns of political correctness. Quasi-autonomous units such as the Ethiopian Valleys Development Authority and the Water Resources Department have been dissolved or stripped of autonomy. So thin is Ethiopian expertise that the death of Zewdie Abate (director of the EVDSA) and the return to academic life of Mesfin Abebe (Minister of Natural Resources Development and Environmental Protection) instantaneously created leadership vacuums. Ethiopia seems perpetually at the very beginning of a Nile learning curve.

Egypt, by contrast, has won the war for expertise and for influence abroad virtually uncontested. Ever since the colonial period, the Ministry of Public Works/Irrigation/Water Resources, containing the venerable Nile Control Department, and, from the mid-1970s, the Water Research Center, has built formidable administrative and engineering infrastructure with remarkable continuity in expert personnel. The water administration is unusual in that it escaped the political struggles that went on inside the Nasserist regime and then in the transition from Nasser to Sadat and from Sadat to Mubarak. For some twenty years the Water Research Center was run by Mahmoud Abou Zeid, who in 1997 became the Minister. Moreover, Egypt, particularly with the advent of the Sadat era and the reentry into the western camp in the Cold War, began to place its water experts and other senior personnel in key UN agencies, the World Bank, the International Monetary Fund, the African Development Bank, and specific bilateral aid agencies. They in turn made Egypt's case for conserving the status quo established through the 1959 agreement and developed the contacts and networks that could stymie any counterstrategies mounted by Ethiopia and other upper basin riparians.

The Sudan, like Egypt, has shown notable continuity in its irrigation and water technocracy, all the more remarkable in that changes in the political regime have been frequent, dramatic, and sometimes violent. A civilian, parliamentary government was toppled by Major-General Ibrahim Abboud in 1958. He in turn was driven from power by civilian protests and demonstrations in 1964. Another civilian interlude was ended by the coup of Ja'afar al-Nimeiri in 1969. His regime was frequently challenged. It underwent a kind of internal Islamist transformation in 1983 and then was driven from power by civilian demonstrations in 1985. From 1986 to 1989 a fragile, elected civilian government was in place, but it was brushed aside in 1989 by Col. Omar Bashir and his erstwhile civilian Islamist ally, Hassan Turabi. Through all this the Sudanese experts in the water sector came through essentially unscathed, and some older figures lived and consulted in Khartoum. One such is Yahya Abdel Mageed, former Minister of Irrigation and the Secretary General of the UN Water Conference at Mar del Plata.

EQUITY AND HARM IN PRACTICE:
SECOND-LEVEL DEALS

Until the 1990s, Nile issues were addressed on a continuous basis by only Egypt and the Sudan. The agreement they signed in 1959 is the sole understanding on the use of the Nile accepted as binding by the two signatories. Other agreements exist, but because they were signed under colonial auspices, one or more signatories do not regard them as binding (Okidi Odidi, 1979). Another agreement, between Egypt and Uganda, negotiated between 1947 and 1949 (see Chapter 7), pertained only to the rules governing the operation of the Owen Falls Dam at Jinja, Uganda, where the Victoria Nile leaves Lake Victoria.

The 1959 agreement superseded that of 1929. Both in essence concern allocation or apportionment of water. The principles underlying the two allocational rules vary significantly. Why that is the case, and what that tells us about the relative weight given to equity and harm in the two instances, is the subject of this section.[10]

In 1925, the most senior British official in the Anglo-Egyptian Sudan, the Sirdar Sir Lee Stack, was assassinated, and the British authorities held the Egyptian government of Prime Minister Sa'ad Zaghloul directly responsible. With a graphic display of fissures in its hegemonic facade, Britain demanded reparations from Egypt and threatened to divert Nile water to the Gezira scheme in the Sudan. The construction of the Sennar dam on the Blue Nile was already under way, and it became clear that some sort of formal understanding between Egypt and the Sudan was necessary. An agreement was signed in 1929 at a time when Great Britain could heavily influence its contents; Egypt enjoyed only nominal independence, and the Sudan was under Anglo-Egyptian control.

The utilizable flow of the Nile was estimated at 52 bcm per annum. Total discharge as measured at Aswan was over 80 bcm per annum. The 30 bcm difference between total and utilizable flow simply flushed into the Mediterranean. It represented the peak of the annual flood that could be neither stored nor used directly in agriculture.

The 1929 agreement stipulated that Egypt would receive on average 48 bcm and the Sudan 4 bcm. This meant Egypt was allocated 92.3 percent and the Sudan 7.7 percent of the utilizable flow. The award thus placed overwhelming emphasis on maintaining Egypt's acquired rights and on avoiding appreciable harm to Egypt's agricultural sector. The small award to the Sudan is nonetheless significant. Until the early twentieth century the Sudan had made no significant use of the Blue Nile or the White Nile for agriculture. It had thus failed to establish any use rights to the river. By definition it could suffer no harm to rights and patterns of water use that did not exist. The award of 7.7 percent of the utilizable flow may have reflected British anger at the assassination of a colonial official, but it also recognized the principle of equitable use by protecting future and, at the time, hypothetical Sudanese claims to put the water to work.

The 1959 agreement, the only one in postcolonial Africa that provides an allocational formula, has a more complex logic. First, the annual mean discharge of the Nile as measured at Aswan, based on the measured flow over the first fifty years of the century, was put at 84 bcm. Because of the planned construction of the Aswan High Dam, the estimated *utilizable* flow of the Nile rose from 52 to 74 bcm. Very little or no water need be allowed to flow to the Mediterranean once the dam and its reservoir were operational. It was estimated that 10 bcm would be lost in the reservoir behind the Aswan High Dam to surface evaporation and to seepage.[11] There was thus a net gain of 22 bcm between 1929 and 1959. The basic allocation of the entire discharge was 55.5 bcm for Egypt and 18.5 bcm for the Sudan.[12]

In Figure 3.2, three allocational formulas are shown. The first is the 1929 settlement, in which the award according to avoiding harm is heavily shaded. The second represents the gross allocations to Egypt and the Sudan in 1959. There we see that Egypt received 75 percent of the gross allocation of 74 bcm and the Sudan 25 percent. There is already a dramatic shift from honoring acquired rights to recognizing equitable use (at a time when the ILC had not even begun to debate the issues). The shift is even more striking, but to my knowledge almost never cited, when one examines the apportionment of the *net* gain of 22 bcm. The Sudan's net gain was 14.5 bcm (18.5 bcm awarded in 1959 minus 4 bcm awarded in 1929) while Egypt's was 7.5 bcm (55.5 bcm awarded in 1959 minus 48 bcm awarded in 1929). From this perspective, two-thirds of the net gain was allocated according to equitable use and one-third according to acquired rights.

There is little doubt that Egypt bargained from strength. The Sudan was no match for Egypt either economically or militarily, and its government was new, untested, and isolated. So what does the settlement tell us? First, because the net gain was additional water heretofore used by no one, Egypt could afford to

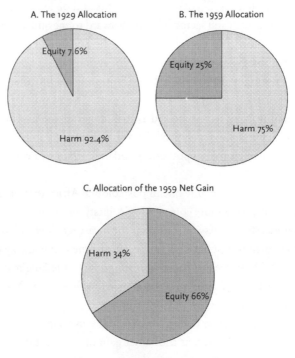

Figure 3.2. Avoiding Appreciable Harm and Allocating Equity
in Egypto-Sudanese Agreements, 1929 and 1959

be magnanimous. It could not invoke appreciable harm as a result of a net allocation that greatly favored the Sudan. But Egypt, in conceding two-thirds of the net gain to the Sudan, went beyond principles codified in other parts of the agreement. It is stated, for instance, that costs for all future projects to reduce losses in the system would be shared on a 50–50 basis, and all net gains in water resulting from such projects would likewise be shared 50–50. Egypt could have but did not insist that the net gain of 22 bcm be shared 50–50.

Moreover, both countries anticipated that other riparians might one day make claims on Nile water outside the framework of the 1959 agreement. In that event, the agreement stipulated that Egypt and the Sudan would negotiate jointly with any other claimants. If any water were to be allotted to another riparian as a result of negotiations, then Egypt and the Sudan would deduct the allotment from their shares in equal parts. The readiness, stated in writing, to entertain such claims underlines the implicit acknowledgment of rights to equitable use of other riparians, all of which had even fewer claims than did the Sudan to the water based on established use. Any such deduction would necessarily cause harm, appreciable or not, to patterns of use put in place by the 1959 agreement.

With the passage of time, the original spirit of the 1959 agreement has been mostly lost.[13] Predictably, many in Egypt and in the Sudan at the time were convinced that the negotiators on behalf of their respective countries had sold out. Egypt had not insisted on a sufficient share of both the gross and net amounts. Many Sudanese felt that 18.5 bcm provided very little of the water the country's future agricultural development would require. Such afterthoughts probably indicate a balanced agreement. The shares, in absolute amounts, have now become sacrosanct, absolute floors below which neither country will go. By the time President Sadat issued his warning in 1979, the Egyptian share had been graven in the stone of Egypt's strategic interests.

It is thus somewhat surprising to find, in 1995, a senior Egyptian diplomat, Deputy Assistant Minister of Foreign Affairs Marwan Badr, laying out a blueprint for Nile basin cooperation that was in tune with the spirit of the 1959 agreement (Badr, 1995).[14] He notes that the other Nile riparians have a growing need for water given the expansion of their populations. The pressures Egypt feels are therefore more than political and reflect real demand that Egypt must take into account. He pointed out: "We are obliged to engage in a process of negotiations that will ineluctably lead us toward the re-definition of quotas. What we are seeking at the present time is to offer them alternatives. Ethiopia or Uganda, for example do not need water but [rather] electricity. We, then, propose the construction of dual use projects; water for us and electricity for them." Far more often, however, Egyptian policy-makers have taken the "quotas" as unalterable and concentrated instead on the "alternatives."

The basic Egyptian position, arduously defended, is that Egypt's 55.5 bcm are absolutely essential to Egyptian agriculture and to its growing population of some 65 million at the turn of the century. Safwat Abd al-Dayem, Secretary-General of Egypt's National Committee for Irrigation and Drainage, was quoted in *Le Progrès Egyptien* (September 15, 1996) as saying: "We have very limited water resources, and they are barely sufficient to cover our needs." A few months later, as we shall see below, Egypt suddenly "found" nearly 5 bcm in its water supply.

THE EGYPTO-ETHIOPIAN CHESS MATCH

Egypt seeks basin-wide cooperation, but only in the sense of helping other riparians develop alternatives to Nile water. Enjoying, by seniority, a near monopoly in expertise on all the engineering and agronomic issues facing basin-wide development, Egypt offers technical assistance in helping other riparians exploit alternatives. These might consist in development of watercourses that do not feed into the Nile, improving rainfed agriculture, and implementing sophisticated water harvesting and water reuse programs. Since the 1980s, Egypt has mobilized annually millions of dollars in direct assistance to other riparians, including Ethiopia.

At the same time, part of the Egyptian strategy has been to insist on data-gathering and extensive studies of specific problems and projects. Egypt's adversaries have come to regard these quests for information as stalling tactics as well as attempts to learn more about the water balances of the upstream riparians. Since the late 1970s, USAID has funded a number of projects in Egypt to improve data-gathering and analysis of the Egyptian irrigation system and of water resources in the Nile basin as a whole. The latter effort has involved close coordination with the U.S. National Weather Service and the utilization of satellite imagery. While all riparians could benefit from the data generated through this program, none has agreed to participate. [15]

The first manifestation of Egyptian strategy came in the mid-1960s. Between 1961 and 1964, the level of Lake Victoria and other equatorial lakes rose dramatically, in the case of Victoria by some 2.5 meters (over eight feet), leading to extensive coastal flooding and displacement of populations (see Chapter 7). The causes were not known, prompting the World Meteorological Organization, with Egyptian backing, to undertake a survey of rainfall in the affected regions of the upper Nile. The project, known as Hydromet (the Hydrometeo-rological Survey), was begun in 1967, and the first phase of catchment survey ended in 1974. The second phase, 1974 to 1981, consisted in the elaboration of a mathematical model of the precipitation and runoff regimes, and data analysis (see Hydromet, 1974 and 1981). The project undertook basic training of techni-cians and technocrats in several of the newly independent riparians. Egypt and the Sudan, while lying outside the upper basin which was the focus of Hy-dromet, were full members while Ethiopia joined only in 1971 as an observer and a skeptical one at that. The headquarters were at Entebbe, Uganda, where Egypt, through the Owen Falls agreement, had considerable influence.[16]

In 1979, university professor Boutros Boutros-Ghali became Minister of State for Foreign Affairs in Egypt. He was an expert on the African state system as well as international law. He made Africa and the Nile basin the focal point of his efforts as Minister of State. He wanted Egypt to present an image of a power that could produce win-win solutions to all riparians. To do this, Egypt had to move beyond its insistence on alternative water resources as the key to cooperation to an emphasis on what I would call multi-good solutions. Mar-wan Badr, in his 1995 interview, stresses this theme.

In essence, under Boutros-Ghali, Egyptian Nile policy was aimed at diffus-ing the focus on water and redefining the problem as one of basin-wide development. Working together, all riparians could focus on transportation, tourism, public health, interriparian investment, and regional security. The single most significant water-related good was power development. Egypt, in conjunction with its then-ally President Mobutu of Zaire, studied the possibil-ity of transmitting excess power generated at the Inga dam on the Zaire River eastward across Africa to the Nile basin, and thence northward across Egypt and Jordan, to Turkey and Europe. The Egyptians thought this could provide Europe with an alternative to potentially dangerous nuclear power generation. All African countries across whose territories the transmission lines would go could tap into this source. In addition, as Badr suggested, Egypt could help countries like Ethiopia develop storage sites in the highlands to supply power to local markets while letting the stored water flow downstream to the Sudan and Egypt. Moreover, it was suggested that the economically weak riparians of the basin would be better able collectively, and implicitly under Egyptian leader-ship, to raise capital whether from multilateral or commercial sources. In 1983, with the nudging of Boutros-Ghali, the Undugu group of Nile riparians was formed. (*Undugu* is a Swahili word for brotherhood.) Ethiopia saw it as another

Egyptian Trojan horse and opposed it from the outset. Other riparians went along with it but without conviction.

Ethiopia effectively scuttled Undugu. Both under the *dergue* and under the post-1991 governments, the development of the power of the Zaire (now Congo) River was seen as an Egyptian maneuver to preempt development of Ethiopia's huge untapped hydropower potential. In the eyes of Ethiopian officials the multi-good game being proposed was loaded with multiple negative outcomes; that is, both Ethiopia's irrigation and hydropower capabilities were likely to remain stunted. The other key actor, Uganda, was locked in internal struggles that led to the ouster of Milton Obote and the advent of Yoweri Museveni. By the time Boutros-Ghali was elected Secretary-General of the United Nations, the initiative had died, and Egypt still had not won official recognition on the part of the upper riparians of its "rights" under the 1959 agreement.

Multilateral and bilateral actors attempted to fill the breach. The Canadian International Development Agency (CIDA), UNDP, FAO, and World Bank all backed, with varying degrees of enthusiasm, a successor organization to Hydromet. In December 1992, Tecconile (Technical Cooperation Committee for the Promotion of Development and Environmental Protection of the Nile Basin) came into being with six member states: Egypt, the Sudan, Rwanda, Tanzania, Uganda, and Zaire. Ethiopia, Kenya, Eritrea, and Burundi were observers. The launching of Tecconile corresponded to the consolidation of power of the new government of Ethiopia, led by Meles Zenawi. The Cold War had ended, Russia had neither the will nor the resources to engage in the Horn of Africa, and the Ethiopian government recognized the necessity of attracting foreign private investment and favorable treatment from the World Bank. In short, Ethiopia was trying to get back into an international arena in which Egypt had become a skillful and respected player. Ethiopia therefore approached Tecconile as an active participant without being, de jure, a full member.

The basic strategies of Egypt and Ethiopia became clear within three years and were formalized at the third ministerial meeting in Arusha, Tanzania, in 1995. Egypt concentrated on the Plan of Action, consisting of some twenty-two projects under five headings: integrated water resources planning, capacity building, training, regional cooperation, and environmental protection. The first three were direct holdovers from the Hydromet project. The initial price tag was estimated at $100 million. In general the Egyptian objective has been to emphasize ways by which the upper basin riparians can make better use of their water resources. One of the prime actors in the CIDA, Aly Shady, along with two

Sudanese coauthors, Ahmad M. Adam and Kamal Ali Mohamed (1994:79), tried to set the parameters for any wider discussion of water sharing: "If all riparian countries sharing the basin water resources come to a consensus to negotiate a basin-wide agreement for water allocation, such an agreement should take into account the technical, social, and economic aspects, the needs, the actual uses, and existing bilateral agreements."

Ethiopia, at Arusha, countered with a call for the elaboration of a framework for cooperation that would have to *precede* any action plan. Action can be taken, the Ethiopians argued, only after we have agreed on a users' code, preferably based on principles of equitable use. The Ethiopians called for the formation of a Panel of Experts (sometimes referred to as D3 in the Arusha proceedings), drawn from the legal and water specialists of the riparian countries, to design the framework. This proposition was the only proposal unanimously adopted by all the riparians present at the Arusha Tecconile meeting (Alemu, 1995:282).[17] Eventually experts in the World Bank dealing with African transboundary water issues, such as Nadendra Sharma, Akhtar Elahi, and David Grey, offered World Bank support in bringing together the Panel of Experts.

The panel was constituted in 1997. The Ethiopians thereby won a significant victory, but skeptics still believed that the experts themselves would not be able to stray from the national lines of the countries from which they came. In other words, rather than avoiding "talking shops," Ethiopia may have prompted the creation of yet another one. The non–water-related disputes that intensified in 1997 and 1998 throughout northeast Africa—Ethiopia versus Eritrea, Uganda and Rwanda's involvement in the Congo rebellion, the continued tensions among Egypt, the Sudan, Ethiopia, and Uganda—set strategic concerns among the Nile riparians that would push any cooperative framework to the end of the queue of national priorities.

One may reify the rival positions in the following terms. Ethiopia took a stand on principle, calling for a code embodying justice and equity. The code, in Ethiopian eyes, relativizes and marginalizes the 1959 agreement, if it does not lead to its outright rejection as a governing document. The code should lay down the principles by which a new allocation can be made. It was doubtless this understanding that explains the great reluctance of Egypt to approve D3 at Arusha. Egypt has offered process, working on a complex set of action items which all, to some extent, require national rather than international measures. The implementation of the action program will require data-gathering, interriparian meetings, training, monitoring, and the joint pursuit of funding. It is a long-term process, requires no consensus on a users' code, leaves intact all existing agreements (of which there is only one of significance), and has as its objective the development of *additional* water supply in the Nile watershed, thereby obviating any new allocational formula.[18]

The other riparians, with lower stakes in the outcome, seek a middle ground. They do not want to antagonize Egypt, and some of the items in the action plan would be of direct benefit to them. At the same time, they resent Egypt's quasi-hegemony and are unhappy with the 1959 agreement. The best role for them, one that Uganda has played with some skill (see Chapter 7), is to try to mediate between Ethiopia and Egypt. The Sudan, a direct beneficiary of the 1959 agreement, almost never breaks rank with Egypt on Nile issues in public, but its senior officials frequently voice private concerns that the water interests of the two countries may be on a collision course. I shall discuss these issues in some detail in Chapter 6.

ENTER THE ENTREPRENEUR

Despite the unfavorable atmospherics in the region, the World Bank and the UNDP continued their role as third-party catalysts to a basin-wide convention. They became the entrepreneurs of cooperation. The question once again arises as to why third parties, distant from the basin itself, would seemingly have higher motivation to achieve agreements than the riparians themselves. Two tentative responses are that professionals in these and other donor organizations are rewarded for their achievements, and neither the organizations themselves nor their professional staff have any obvious stake in the status quo. Because of their experience, they understand better than some of the riparians the benefits of cooperation as well as the means to bring it about.

Between 1997 and 2001, considerable progress was made in designing the structures and institutions of a new regime in the basin. Tecconile has been replaced by the Nile Basin Initiative.[19] Its Secretariat has taken over the Tecconile headquarters at Entebbe, and a Tanzanian, Meraji Msuya, was appointed executive director in 1999. Work is proceeding on a cooperative framework, drafted by the panel of experts provided for under D3. The objective is to establish a Nile River Basin Commission grouping all riparians. To date, Eritrea has not been a party to these efforts.

The principles proposed in the draft of December 10, 1999, but not yet adopted by the member states, reflect the Helsinki Rules and the 1997 ILC document. It is significant that equitable utilization is given prominence, perhaps even precedence, although, as in the ILC document, the prevention of "significant harm" is also stressed. This draft begs the familiar questions of the criteria by which to assess rival or contradictory principles and claims. One set of recommendations, based on D3 and called Output 2, lays heavy responsibility for assessment and recommendations on Strategic Advisory Committees, composed of water experts from each member state. The committees are urged to develop a decision support system "for the analysis and modeling of options." It is not hard to imagine how the committees might become heavily politicized if the data and interpretations they produce feed into allocational decisions.

The other main institutional components include the Council of Water Resources Ministers, which is the supreme authority of the proposed Nile River Basin Commission and reports to the Conference of Heads of State and Government. There is to be a Technical Advisory Committee, with two expert members from each member state, which advises the Council of Ministers on technical matters, the modification or introduction of new principles and procedures, and the feasibility of proposed projects. It in turn is empowered to appoint and define the tasks of Strategic Advisory Committees.

The draft principles endorse the concept of subsidiarity, that is, carrying out as much as possible at the lowest jurisdictional level. Toward this end, each member state is to establish a National Focal Point Institution to work toward harmonizing national policies with the basin-wide principles. The draft recognizes the utility of existing sub-basin agreements and institutions, although the Egyptians and Sudanese expressed reservations about the proposal to reconcile the operating principles of those sub-basin units with the Nile Basin Initiative. The draft refers to three interlocking sub-basin units: Egypt and the Sudan, bound already by the 1959 agreement, linked, on the one hand, with the Eastern Nile group of Ethiopia and Eritrea, and, on the other, with the Southern Nile group of the Democratic Republic of the Congo, Uganda, Kenya, Rwanda, Burundi, and Tanzania.

The Nile River Basin Commission will enjoy international legal personality, the capacity to enter into legal agreements, incur obligations, receive donations, and sue or be sued in its own name.

The donor community will be represented through the International Consortium for Cooperation on the Nile (ICCON). It will be the provider of carrots for the member states[20] and presumably will help fund a number of win-win projects, held over and amplified from the Tecconile action plan. These will involve training, infrastructure development, trade, tourism, industrial promotion, environmental protection, afforestation and watershed management, transportation, and hydropower development and pooling.

Despite this progress, Ethiopia insists on a new set of allocational and use principles that could call into question certain aspects of the 1959 status quo, if not the legality of the 1959 agreement itself. Insofar as projects are concerned, one informed expert who shall remain anonymous, wrote me that win-win for Ethiopia means acceptance within the Nile Basin Initiative of a "couple of multi-purpose hydropower projects with significant consumptive water use components."

UNILATERALISM IN THE BASIN

The task of third parties in mediating Egypto-Ethiopian differences became somewhat more urgent since President Mubarak announced, in the fall of 1996, Egypt's intention to use Nile water to develop new agricultural lands in the string of oases in Egypt lying to the west of the Nile.[21] The announcement, amidst great fanfare, effectively put an end to the few positive steps the two countries had taken to explore their differences over the use of the Nile.

Soon after the Transitional Government of Ethiopia had come to power, it signed with the Sudan a Declaration of Peace and Friendship (December 23, 1991), implicitly recognizing thereby the Sudan's role in helping the Eritrean, Tigrean, and Oromo liberation fronts to overcome militarily the armed forces of the *dergue*. Several articles of the declaration dealt with principles for the use of the Nile waters. They referred to the Nile as "a common resource of the co-basin states" and pledged coordinated management of their joint watersheds, especially implementing measures to reduce erosion and sediment runoff in the upper, Ethiopian portions. No mention was made of the 1959 agreement, and the two sides pledged to work toward the establishment of a Nile basin organization representing "the interests of all riparian countries with their universal consent."

A similar agreement was signed on July 1, 1993, between the Transitional Government of Ethiopia and Egypt, entitled "Framework for General Cooperation Between Ethiopia and the Arab Republic of Egypt." Five articles dealt with Nile questions, endorsing principles of mutual consent, periodic consultation, and application of recognized principles of international law. The only principle actually cited was the avoidance of appreciable harm. No mention was made of equitable use. By the same token, no mention was made of the 1959 agreement. Both parties pledged to work toward "effective cooperation among the countries of the Nile basin for the promotion of common interests in the development of the basin" (Alemu, 1995:22). In both agreements, nobody conceded very much, although the endorsement of the appreciable harm principle, in the absence of reference to any others, could be seen as a one-sided commitment on the part of Ethiopia to acknowledge and protect Egyptian acquired rights. These agreements can be considered as framework accords (Chayes and Chayes, 1995), designed to build confidence and shape more focused and specific objectives of formal cooperation. Unfortunately, both agreements failed utterly to serve that or any purpose.

The first bone of contention manifested itself with respect to projects under the aegis of the Ethiopian Relief and Social Rehabilitation Fund, set up shortly after the advent of the transitional government in the early 1990s. One objective of this fund was to settle demobilized military personnel on agricultural schemes. The project called for developing some 10,000 hectares of farmland, one-third of which would be irrigated with water from the Abbay and Teccaze watersheds. The total water requirement would have been about 37 mcm per year or about 0.004 percent of the average discharge of Nile. The World Bank approved $70 million to fund all of the proposed agricultural projects, but, according to its own rules (see Chapter 1) it and Ethiopia had to notify Egypt and the Sudan of the proposed actions. Egypt never vetoed the project but continuously called for more information and study. Years went by until 1997, when the World Bank finally decided that no appreciable harm could be involved and that it could disburse the funds. Those years confirmed all the new Ethiopian government's worst fears about Egyptian intentions.

The second incident came in the wake of the assassination attempt on President Mubarak of Egypt in Addis Ababa in June 1995. The Sudan was accused by Egypt and Ethiopia of having allowed, if not sponsored, the attempt, and of harboring the suspected would-be assassins. In the ensuing war of words, the Sudan threatened Egypt with impeding the flow of the Nile (a hollow threat) and suspended participation in the quarterly meetings of the Permanent Joint Technical Commission on the Nile (real and significant). Egypt's Foreign Minister, Amr Musa, warned that any tampering with the Nile would result in war. Ironically, the incident did nothing to improve Egypto-Ethiopian relations as Egypt criticized Ethiopian security both for allowing the assassination attempt to take place and for poor follow-up in the subsequent investigation. By the fall of 1995, therefore, the two framework agreements had been superseded by events on the ground.

I held conversations in the fall of 1995 and the spring of 1996 with very highly placed Ethiopian policy-makers in the Foreign Ministry and in the Prime Minister's office. They made it clear that they believed that the Nile waters issue amounted to a major structural problem between the two countries and, by that token, Ethiopia's major foreign policy challenge. They had hoped, they told me, that the 1993 agreement would signal Egypt's willingness to accommodate the interests of a new, friendly but economically weak neighbor in the basin. Three years later, they felt that Egypt had been at best politely obstructionist, and, they suggested, in the long term interested in keeping Ethiopia economically weak and diplomatically isolated.

The announcement in Egypt of the New Valley project came as a major surprise to all who study the Nile and Egyptian policy. In the past, even under Nasser, there had been talk of developing the oases using the aquifer that underlies them, but not much had come of these schemes. During the Sadat era, with the filling of the reservoir of the Aswan High Dam, a spillway had been excavated at Toshka midway along the west bank of the reservoir to drain off excess floodwaters in the very occasional years when the dam and the reservoir might operate at above-capacity levels (for a critical assessment of Toshka see Whittington and Guariso, 1983). Under the Mubarak administration, no further mention had been made of significant agricultural development in the western oases. Rather, the talk and the action had centered on the development of the Sinai with water delivered by the Peace Canal; hence the surprise, and in some quarters the dismay, associated with the new scheme.

The core of the scheme lies in building a huge pumping station and canal, beginning at Toshka, to siphon off water stored in the reservoir and bring it by surface to the southernmost oases, 70 km away. There, some 200,000 hectares would be brought under irrigated cultivation, requiring over 5 bcm per year (or nearly 10 percent of Egypt's allotment under the 1959 agreement). Egypt has insisted that these 5 bcm can be found within its existing share (hence the reference above to slack in the system) by reutilizing drainage water and by reducing the area under water-intensive crops, such as rice. Eventually 7 million new inhabitants would move out of the crowded confines of the Nile Valley and into the oases. Initial costs for the pumping station and the canal were put at $2 billion, but total costs for twenty years of development and resettlement were estimated at nearly $100 billion (see Marcus, 1997; Butter, 1998).

The Ethiopians were furious but more or less bit their official tongues during the Nile 2002 conference hosted in Addis Ababa in February 1997. On March 20 a letter of protest was sent by Ethiopia's Foreign Minister, Seyoum Mesfin, to Egypt's, with copies to James Wolfensohn, president of the World Bank, Kofi Annan, Secretary-General of the UN, and Salim Ahmed Salim, Secretary-General of the OAU. The letter stated: "Ethiopia wishes to be on record as having made it unambiguously clear that it will not allow its share to the Nile waters to be affected by a *fait accompli* such as the Toshka project, regarding which it was neither consulted nor alerted." A year later, at a meeting of the OAU in Addis Ababa, Deputy Foreign Minister Tekeda Alemu called for scrapping the 1959 agreement (Reuters, February 27, 1998).

Coming from an upstream riparian, this protest against downstream diversions is not self-explanatory. In 1979 Egypt's Minister of Irrigation, Muhammad Samaha, responded (more calmly than had President Sadat) to Ethiopian protests over the Peace Canal project, stating that downstream, or what he called "estuary," riparians are under no obligation to notify upstream riparians of water development projects unless they were to result in flooding of a portion of an upstream riparian's territory (see Abdul Mohsen, 1980). That is not an unreasonable position.

The Ethiopian protests, both in 1979 and in 1997, are founded on three principles of what Ethiopian officials regard as equitable use. First, if there is slack in the Egyptian allocation under the 1959 agreement, then that slack should be used to accommodate *some* of the water needs of the upper basin riparians. By implication, if the Sudan, for example, is able to run its agricultural sector with less than its 18.5 bcm allotment under the 1959 agreement, the surplus should be used to meet other riparians' needs.

The second principle is that downstream actions can cause appreciable harm upstream. Ethiopia used the term *fait accompli* advisedly. The creation, de novo, of projects that use significant amounts of water may, and probably will, become the basis for asserting new acquired rights founded in established use. Egypt's action in the New Valley (or in the Sinai through the Peace Canal), in Ethiopia's view, preempts Ethiopia's right to harness its Nile water. If the principle of first in time, first in right prevails, then Ethiopia will have to forgo projects of its own in order to protect Egypt's use rights in the New Valley or in the Sinai. Ethiopia will suffer appreciable harm in order not to cause harm to Egypt.[22]

Third, unilateralism is to be condemned wherever and whenever in a watershed it may occur. If one endorses the notion of the community of users and the right of all users to be informed of and to acquiesce in proposed changes in basic use patterns on the part of any riparian, then Egypt was under obligation to inform all riparians of its plans for the New Valley and for the Sinai, and to seek their acceptance.

In tit-for-tat fashion, Ethiopia is moving ahead unilaterally with plans for a major hydroelectric project on the Teccaze River (see Chapter 5).[23] In the current era there are also possibilities of attracting private capital to develop power delivery systems. Far more than irrigation and farming projects, power projects are bankable with fairly short payback periods. There are thus financing opportunities that in an earlier, socialist era Ethiopia would not have contemplated nor had available. The country is also proceeding with hundreds of microdam projects that do not require sophisticated engineering or international financing (see Chapter 4). Cumulatively, such projects can utilize significant amounts of water (see Waterbury and Whittington, 1998).[24]

UPPER BASIN UNIONISM?

All upstream riparians in the Nile basin, including the Sudan, share varying degrees of suspicion toward Egypt and Egyptian motives in seeking cooperative understandings (see, for example, Ochieng, 1996). It seemingly follows that Ethiopia could mobilize these fears and occasional resentments into an alliance of upper basin riparians. The British in fact tried to do just that from 1959 to 1961, as Egypt and the Soviet Union jointly pursued the Aswan High Dam project at the expense of the upper basin. So far, Ethiopia has not tried to emulate the British. Highly placed Ethiopian officials have stressed that the major confrontation is between Ethiopia and Egypt. The other riparians, having much smaller stakes in the game, much more pressing problems than water, and a keen awareness of the costs involved in antagonizing Egypt, will not commit to any coalition of upper basin interests before seeing how the Ethiopian-Egyptian wrestling match is likely to come out.[25]

One issue might elicit a collective response on the part of some of the upper basin riparians, but it is not directly related to Ethiopian interests. Were Egypt and the Sudan to complete the Jonglei Canal (or what is known as Jonglei I), and, even more, if the two countries were to pursue excavation of a second canal (Jonglei II), they might well want to use Lake Victoria as the primary storage site for the water that would go through the canals during the dry season in the Sudd swamps of the southern Sudan. If that in turn entailed raising the level of Lake Victoria with consequent flooding of coastal areas and disruption of ports, shipping, and fishing, the Victoria basin states (Tanzania, Kenya, Rwanda, and Uganda) might collectively reject the Egypto-Sudanese project or collectively seek substantial compensation, if not annual rents.

SLACK IN THE SYSTEM

The 1959 agreement between Egypt and the Sudan, with its allocation formula based on absolute amounts of water, set in stone a zero-sum game. Any abstraction of water upstream meant an automatic loss downstream. All riparians have an interest in transforming this game from zero-sum to variable-sum. The crucial question is whether or not the 1959 agreement is to be part of that transformation, that is, rewritten if not scrapped, or whether the variable-sum game is to involve only new sources of water over and above the 1959 allocation. Egypt wants the latter and Ethiopia the former. The outcome will not be determined by principles of international law, but rather by the relative power of the protagonists and of their allies.

Nonetheless, all parties may want to focus on the generic categories of "slack," available to them for bargaining purposes and for emphasizing variable-sum outcomes. By way of example, let us take Egypt's real supply of water. There are 55.5 bcm annually as the result of the 1959 agreement. At least 6 bcm in water used upstream in Egypt drain back into the Nile downstream, and in reutilized drainage water in the delta. There are probably 5 bcm in water that the Sudan does not use currently under its share from the 1959 agreement, and perhaps as much as 10 bcm in virtual water embodied in food imports. This means that Egypt's effective water supply is around 76 bcm per annum. Its *needs,* officially at least, are far less.

The New Valley is conceived of primarily as an agricultural scheme, but we know that in the Sinai Egypt is contemplating its own Silicon Valley. New settlements based on high-tech industries whose products are destined for export may absorb as many people and use far less water than the New Valley. In terms of Egypt's economic future the Silicon Valley approach is probably superior to the New Valley.

Likewise, as stressed in Chapters 4 and 5, Ethiopia's agricultural future may lie more in the highlands under rainfed conditions than in the lowlands, far from markets and reliant on extensive irrigated perimeters. Hydropower and rainfed agriculture (supplemented by tube wells and pumped groundwater) may make better sense for Ethiopia than trying to replicate the costly and currently malfunctioning Gezira-Managil scheme of the Sudan.

Every riparian can inventory its actual and potential supply in similar terms. When we look at the basin as a whole, we may see four kinds of variables that can create slack:

• virtual water through the importation of agricultural produce;
• technological and pricing innovations to increase efficiency in water use (doing more with less, or more with the same);
• rainwater harvesting and watershed management; and
• varying degrees of drainage of the Sudd swamps in the southern Sudan.

Pursuing any or all of these opportunities to create slack bear costs, and some may be financially and/or environmentally unacceptable. Still, they redefine the parameters of the stakeholders in ways that should make the game more tractable and less potentially destructive.

SUMMARY CONCLUSIONS

I have examined three cycles of the game for control of Nile resources and real estate. In the first, rival colonial powers were the main actors, and they operated through local agents of varying hues and shapes—rulers, cotton interests, traders, tribes, governors, and religious sects. The game was over protecting or extending colonial real estate. Local interests in this first round of a three-level game were much more involved in the process, mainly because the state system of northeast Africa remained in embryo, with only Egypt and Ethiopia having any historical claims to statehood.

In the second round, corresponding to the collapse of empire and the emergence of the Cold War, the northeast African state system came into being. The first-level players were the two Great Powers and their state allies. Local interests of whatever stripe, outside the military, had no meaningful role in the game.[26] Strategic interests at all levels drove policy.

The Cold War round yielded one major agreement, that of 1959 between Egypt and the Sudan. Both countries had the same first-level patron, the Soviet Union. Before equitable use had a name in international law, Egypt went far to recognize it de facto. The allocation of the net gain in water between the 1929 and the 1959 agreements was heavily in favor of the Sudan and not based on established use but rather potential use. Egypt did not have to make this concession. As the years went by, its spirit was forgotten in Egypt, although Marwan Badr briefly invoked it in 1995. How interesting it would be if Egypt offered to finance projects to enhance water supply on a 50–50 basis with all riparians while conceding two-thirds of the net gain in supply to the upper basin states.

One major lesson from the first round of the great game remains valid today. The international community, including major powers, multinational corporations, commercial banks, and international financial institutions, will pay attention first and foremost to the economic and strategic assets of any particular riparian. In this sense, Ethiopia's struggle with Egypt is somewhat quixotic. It is not (yet) economically girded for the struggle. Hence its stance on principle and its righteous indignation over Egyptian unilateralism could win it sympathy but little else. The high moral ground may be in Ethiopia, but the fat contracts are in the New Valley.

The end of the Cold War has radically altered the nature of the game. For good or ill, the highest stakes today are held by the second-level national players. The stakes for great(er) powers have been substantially diminished. That fact, I argue, has been masked by cataclysmic events that focused attention on northeast Africa as the Cold War wound down. Periodic famine in the Sudan and Ethiopia, the collapse of the Somali state and international intervention in its civil war, the genocide in Rwanda, and the collapse of the Zairean state elicited a western commitment of UN forces and emergency aid that has cost billions of dollars and, so far, has not yielded protection against either famine or genocide.

In light of the Somali and Rwandan crises, the then-Administrator of USAID, Brian Atwood, and then-U.S. Vice President Al Gore launched the Greater Horn Initiative in 1994–95 to move beyond emergency interventions of all kinds and to try to deal with the root causes of famine and conflict. The initiative had lost all visibility by 1998 as Eritrea and Ethiopia entered into hostilities, and as the Congo became the theater for confrontation among at least four African states. The southern Sudan was once again wracked by famine as the international community resorted to the usual emergency shipments of food and medicine.

I had speculated, at least to myself, that the objectives set under the Greater Horn Initiative might reveal to U.S. policy-makers some tradeoffs in U.S. policy objectives in the Nile basin. That is, if economic development and political stability were seen as the keys to avoiding civil war and famine in the future, and if economic development in the poor upper basin states hinges initially on stabilizing and increasing agricultural production, then the U.S. might have some reason to encourage Egypt to be more forthcoming in accommodating the water needs of the upstreamers.

The simple truth is, however, that U.S. interests in the Horn are in no way vital to national security and will be visible only episodically. The U.S. and its allies have no significant economic interests in the region. By contrast, U.S. interests in Egypt are far more visceral (as they were for Great Britain). Egypt has been a steady partner in the search for a lasting peace between Israel and its Arab neighbors. It has successfully represented itself as a bastion against Islamic extremism, which, it is suggested, if allowed to spread, could undermine U.S. interests everywhere in the region, above all in Saudi Arabia and the Persian

Gulf. Ethiopia, let alone the other upper basin riparians, can command no such attention in U.S. foreign policy circles, and there are no other external powers to whom the riparians can turn.

Finally, we should expect to see the gradual emergence of a new set of third-level interests in the basin. The increasingly intense use of Nile water resources for agriculture, power, and industry is itself creating local economic interests that one day may have a direct role in the policy process. Examples are the industries and their owners that have sprung up in the Jinja-Kampala corridor in Uganda (interests badly bruised under Idi Amin but today reviving), dependent upon the power generated at the Owen Falls Dam. In Egypt, commercial farmers have a big stake in the reclaimed irrigated perimeters to the west and the east of the delta, and new farming interests may develop in the Sinai and in the New Valley. Veterans groups and refugees in Ethiopia are dependent on irrigated schemes and may in the future acquire some political weight. In short, water policies that had been determined by narrow policy elites in light of national strategic concerns may increasingly be influenced by the new economic interests that were created, intentionally or not, by those policies.

Chapter 4 Food Security in Ethiopia

National strategic interests drive the policy-making and goal-setting among the governments of the Nile riparians. In Ethiopia, the single most important strategic interest is striving to attain food security in a chronically famine-prone region. The quest, as we shall see, may be chimerical, or at least ill conceived, but that does not diminish its enormous saliency in Ethiopian policy-making.

Since the early 1970s, Ethiopia, and other parts of the Horn of Africa, have thrice experienced severe drought and, in the first two episodes, widespread famine. In the first two instances the production and consumption crises contributed to the downfall of the political regimes in place. If for no other reason than their own survival, political leaders in Ethiopia cannot ignore this history. In nearly every respect little has been done in Ethiopia to reduce the risk of drought-induced famine. The road, railroad, and airport networks that might allow the movement of food staples from surplus areas and ports to food-deficient areas have witnessed little improvement in the past decades, and, because of prolonged civil war, have in some places deteriorated. The supply of electrical power throughout the country, the springboard to nonagricultural production, is still geographically limited. The economy as a whole remains heavily dependent upon the agricultural sector, so that when it suffers all the economy suffers with it. Yet the conditions under which agricultural and livestock production take place have, if anything, become more difficult in recent decades. Deforestation, and, in general, the removal of all forms of groundcover, have increased with growing rural populations. Soil erosion and degradation in the densely inhabited highlands has continued apace.

Clearly there is a need for aggressive public policy responses to these challenges, but it is by no means obvious what the most appropriate policy responses are. There is a natural tendency to seek a fix through big, multipurpose projects to harness water for irrigation and to generate hydroelectricity. If rain is both geographically and seasonally unpredictable, as it is, then the superficially obvious answer is to capture and store rainfall, when abundant, for use when it is scarce. The difficulties inherent in this strategy will be explored in Chapter 5. In this chapter, I will treat the more difficult and painstaking strategy of small-scale, locally adapted policies to help rural populations augment production and reduce risks of production collapse. These policies can be implemented while the structural transformation of the economy to a nonagricultural footing gradually takes place.

BASIC PARAMETERS OF THE CHALLENGE

Ethiopia is a big and very rugged country. Size and mountainous terrain present formidable obstacles to infrastructure development. Moreover, there are relatively few large urban centers. If electrical power can help transform the lives of average Ethiopians, then it will have to be delivered to the thousands of scattered villages in which most Ethiopians live.

At the beginning of the 1990s, 88 percent of the population was classified as rural, and the agricultural/livestock sector contributed 48–55 percent of GDP and, in value, 90 percent of all exports. Yet basic food grain production, at 6–7 million metric tons per year, fell well short (ca. 15 percent short) of basic requirements (Abate, 1991:59; Donovan, 1996). Ethiopia, by any measure, is one of the half-dozen poorest countries in the world. At least 40 percent of the population lives in poverty, and per capita income (purchasing power parity at 1995 prices) was, in the mid-1990s, about $450 (Egypt's, by comparison, was $3,820: *World Development Report,* 1997: statistical appendix). In terms of caloric intake more than half the population survives on 1,700 calories or less per day.

Ethiopia's surface is, at 1,087,816 square kilometers, about twice the size of France. It has only 24,000 kilometers of roads of all types, of which merely 3,500 are asphalted. By way of comparison, France, with half the surface size of Ethiopia, has 812,000 kilometers of roads (i.e., thirty-four times Ethiopia's). During the rainy season from June to October much of Ethiopia's skimpy road grid is washed away or is otherwise impassable.

The national power grid runs essentially along a north-south axis from now-independent Eritrea and the port of Assab, parallel to the Rift Valley through Addis Ababa to the southern Rift lake area. There are spurs to the northeastern cities of Dire Dawa and Harrar, following the rail line from Djibouti, and shorter spurs westward to Lake Tana and to the towns on the escarpment separating the Blue Nile watershed from the Akabo-Baro watershed. The majority of all Ethiopians do not have access to electricity.

MALTHUS REDUX?

Allan Hoben (1995) has written an interesting critique of the Malthusianism inherent in the foreign assistance discourse concerning Ethiopia.[1] He argues that the empirical evidence does not support the common assumption that population growth among animals and humans is causing massive deforestation and soil erosion that will lead to the permanent collapse of agricultural production, or, at best, become an insurmountable obstacle to increased production. There is some evidence in Ethiopia (see S. Abate, 1994) and in other developing countries that as dense rural populations mine their resources beyond sustainable levels, they take defensive, protective actions such as reforestation and watershed rehabilitation to halt the loss of biomass and the soil erosion process. It is also an incontrovertible fact that over millions of years the Ethiopian highlands have sent downstream the sediment load that has built Egypt's Nile delta.[2]

The question, then, is whether anything going on now is unusual, and, if not, how can one lend credence to the predominant Malthusian developmental gloom?

Although throughout history Ethiopia has, in different places and at different times, been deforested and reforested, it appears clear that some critical thresholds in land use under relatively unchanging practices and technologies of cultivation and husbandry have been approached or exceeded. The Ethiopian population, which in 1900 numbered about 11 million, has been growing rapidly since World War II, at rates around 3 percent per annum, and it has exceeded sixty million with no slowing yet in sight. As already noted, the great majority of that population lives in the countryside and draws its livelihood from the land. There is no reason to suppose that the impact of a quintupling of the population over the last one hundred years has been environmentally benign.

The most authoritative and extensive surveys of the agricultural ecology of Ethiopia were carried out by the FAO (1986) and by the UNDP/FAO (1988), a few years after the last great famine and a few before the collapse of the socialist *dergue,* led by Menguistu Haile Mariam. Their assessments concluded that nearly 20 percent of the highlands, especially in the drought-prone regions of Wello, Eritrea (still part of Ethiopia at the time of the study), and Tigray, had experienced such severe soil erosion that topsoil had been reduced to about 20 centimeters, below which sustainable cultivation is very difficult. It was estimated that extensive areas of the central highlands would reach the same state within two decades. On average, erosion carries away 100–120 tons of sediment per hectare each year. That means close to 2 billion tons annually, although 90 percent of that is redeposited elsewhere in Ethiopia (Constable, 1986). Even in "normal" years reliable cultivation under current practices is problematic.

It is partially true that the extent of deforestation has been exaggerated and the extent and nature of afforestation minimized. One frequently reads that 40 percent of Ethiopia's surface was covered in forests a century ago (a probable overestimate) and now only 3 percent is (accurate). As has been seen in the southwestern Illubabor region over the past thirty years, the impact of human settlement and cultivation has been variable and determined by a host of factors including changes in property rights, changes in relative prices, and soil depletion. Solomon Abate estimates that from 1900 to 1960, deforestation of the Akabo-Baro watershed took place at a rate of about 1 percent of the forested area per year. Soil exhaustion and a move into planting coffee trees led to reforestation and increased fallows at a rate of about 1.3 percent per year until 1975. Then the weight of administratively determined coffee prices set by the *dergue,* the introduction of communal landholding, and the encouragement of food crops led to a reconversion to farmland at a rate of 0.75 percent per year. On top of this, the 1983–85 drought and famine led to the influx, deliberately planned, of several hundred thousand "refugees" from the famine-stricken highland provinces. The refugees did not have easy access to the land, and many made a living from marketing forest products, especially charcoal.

As was the case for soil erosion, the logic of large, poor, rural populations mining their environs seems implacable. Major Cheesman's voyages up and down the Blue Nile and its tributaries in the 1930s produced descriptions of dense, nearly impenetrable forests, which today are gone except in the steepest and most inaccessible gorges. Similar descriptions were written by Wilfred Thesiger (1987) about the Awash Valley, today the seat of Ethiopia's most advanced, large-scale irrigation schemes for sugar cane and cotton. The extraordinary rates at which the reservoirs on the Awash River and on the Sudanese portion of the Blue Nile and Atbara have filled with sediment in the past three decades is hard evidence of the combined impact of extensive cultivation and clearing of groundcover in the highlands.

DROUGHT AND FAMINE

The drought and famine zones of Ethiopia have been well established throughout the country's recorded history. They run in an arc from the center-north (Tigray, Aseb, Wello, and Eritrea) to the center-east (Harargue, Dire Dawa) then to the southeast (Bale, Ogaden) and to the deep south bordering Lake Turkana and Kenya (Borena and Omo South). Famine has struck these regions frequently, but food shortages have never been caused by drought alone. Between 1540 and 1742 there were ten major famines in which local wars exacerbated production failures. Between 1888 and 1892 there was a large famine, intensified by the decimation of livestock by rinderpest. In the twentieth century the search by the central government for additional tax revenues, including land, animal, and hut taxes, increased the burden on peasants living at the edge of subsistence, as did church tithes and the provision of free labor to landlords (see Webb et al., 1992; Kloos, 1991). In recent decades agricultural income and head taxes have yielded on the order of 15–30 percent of total agricultural income (Diriba, 1995).

In sum, many of Ethiopia's famines confirm Amartya Sen's general proposition that famines are the result not so much of production failures and scarcity as they are of a collapse in buying power of the most vulnerable rural inhabitants (Sen, 1981). That was in large measure the case in 1972–73, when 200,000 Ethiopians died in the northern provinces, especially in Tigray and Wello. That famine destabilized the empire and the emperor, and set the stage for the university students and the radical officers of the armed forces to challenge successfully Hailie Selassie's order. The far more devastating famine of 1983–85, however, can be attributed much more directly to severe drought and production failures. More than 1 million Ethiopians may have died, again with Wello most severely affected. But fourteen regions and nearly 11 million people were at risk. Total production of cereals and pulses fell by 18 percent in 1983–84 and

25 percent in 1984–85. Even normally surplus regions, like Arsi, Shewa, and Gojjam, registered sharp drops in production. Food aid, equivalent to 28 percent of national production, was brought in but, because of the civil war raging in the north, was not easily delivered to the most affected populations. Dejene Aredo concluded that food was not available at *any* price (Aredo, 1989:61–62). As in the previous episode, the now-socialist military regime of Ethiopia paid the price, losing militarily to the combined forces of Tigrayan and Eritrean challengers in 1991.

By the mid-1990s, the new, post-Menguistu regime, led by Meles Zenawi, declared that famine was a thing of the past, but in 1998 drought-induced production failures began to occur in the traditional northeastern famine zone, and then, in 1999 severe drought and incipient famine declared themselves in the southeast. The Ethiopian government admitted that 8 million of its citizens were at risk.

At the same time, in the spring of 2000, Ethiopia launched a major and successful offensive along its border with Eritrea, bringing to an end, perhaps, a two-year-long war. The government's popularity, generated by military success, may deflect public attention from failure in agricultural production. The war has also meant that a more agile international community cannot ship food relief to Ethiopia through the Eritrean port of Assab but must rely instead on Djibouti and Mogadishu.

THE PRODUCTIVE BASE

About one-third of Ethiopia's surface is cultivable, or some 33 million hectares. Estimates of what is actually cultivated range from 6 million to 26 million hectares, reflecting the distinction between pastoral use and farming, and the resort to fallows (ADE, 1996). There are five major agro-climatic zones:

a. the highland cereal zone,
b. the highland *ensete*[3]—root crop zone,
c. western, moist, lowland zone,
d. eastern, moist, lowland zone, and
e. the dry lowland zone.

Most of the population is concentrated in zones a and b.

In absolute terms total food production has been stable. In per capita terms it has been steadily declining, even when adding in the imported commodities and food aid (Figure 4.1). The reasons are not hard to discern. In the highlands

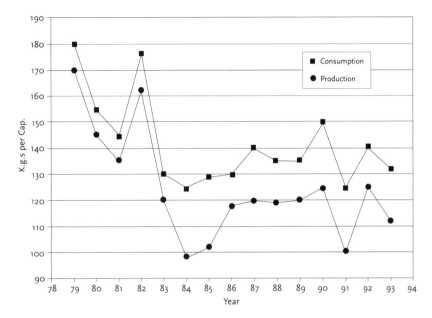

Figure 4.1. Grain Consumption in Ethiopia Per Capita, 1979–93
Source: Julie Howard et al. "Toward Increased Domestic Cereals Production in Ethiopia,"
Working Paper 3, Food Security Project, Ministry of Economic Development and
Cooperation, Addis Ababa, 1995.

it is rare to see any use of mechanical traction or pumping. Tractors are to be
found only on the large state farms in the Awash Valley and on the commercial
farms in the lowlands along the border with the Sudan. Although in recent
years fertilizer imports have increased fourfold, prior to the mid-1990s much
smaller amounts were reserved overwhelmingly for state farms that represented
only 4 percent of the cultivated surface (Abate, 1991:59). The most advanced
technology in the highlands remains the traditional ox-drawn plows.

Emperor Haile Selassie was reluctant to tamper with the existing, unequal
(feudal is the term historians of Ethiopia often employ) structures of landholding for fear of eroding his own political support. Rather, he saw the way out of
Ethiopia's low production trap as lying in large-scale irrigation schemes in hitherto lightly inhabited and primitively cultivated regions. The first choice was
the Awash Valley, developed in the 1950s, draining a portion of the Rift Valley
in a northerly direction toward Djibouti (see Chapter 5). Afar pastoralists were
displaced, sometimes with devastating consequences, but they were a nut far
easier to crack than the property structures of the highlands.

The Marxist regime of Menguistu Haile Mariam was more ambitious. It decreed all agricultural lands to be in the public domain and dispossessed the landowning class. Although peasants were not given title to land, they were given usufruct rights. Operational units nonetheless declined from an average of 1.3 hectares per farm family in 1978–79 to 0.8 hectares in 1988–89 (ADE, 1996). At current levels of technology a subsistence holding is 3.7 hectares (Diriba, 1995:343). So the land reform did not solve the production challenge or alleviate rural poverty. Projected state farms, on the Soviet model, particularly in the moist Southwestern lowlands, were seen as the solution to enhanced production. The philosophy, such as it was, is embodied in the Ten-Year Perspective Plan, 1984/85–1994/95. It was drafted in the midst of the drought/famine. It called for expansion of the cultivated surface from 6.9 to 8.2 million hectares, doubling the production of food staples from 6.6 to 12.2 million tons, and raising daily caloric intake from 1,700 to 2,000. The means identified by the *dergue* were tractor stations, sale of draft animals on credit, and small- and large-scale irrigation schemes (Wolde-Mariam, 1989).

No sooner had the ink dried on this plan than the full extent of the 1983–85 catastrophe became known, forcing the *dergue* to rethink its approach. The FAO and UNDP surveys mentioned above were part of this, as were a number of consulting engineers' reports on hydropower development (see Chapter 5). Greater attention was paid to small-scale agriculture and to rehabilitation of watersheds and degraded lands. At the same time, the regime, with the support of its outside consultants, fastened upon the need to relocate population from drought-prone highland regions to the moist lowlands: "There seems little alternative but to redistribute the population if alternative sources of income outside agriculture [are not found] in the short to medium term" (UNDP/FAO, 1988:61–62). Despite the extent of the crisis, the *dergue* did not lighten the indirect tax burden on agricultural producers. The state-owned Agricultural Marketing Corporation held purchase prices for most commodities at about one-third the import parity price (ADE, 1996).[4]

Ethiopian "traditional" agricultural has not evolved technologically. The growing season is determined by the length of the short, *belg,* rains in March and April, and the long, *mehr,* rains of the Ethiopian winter months, July through September. Traditional and supplemental irrigation is practiced on only about 65,000 hectares, or less than 1 percent of the cultivated surface. There are five major staple crops. Three, with relatively long maturation periods, are dependent on the very fickle short rains at the beginning of their growing periods: maize, millet, and sorghum (requiring 120–210 day growing periods). *Teff* (the indigenous highlands "wheat") and barley have shorter maturation periods, in the range of 90–120 days, and depend on the long rains. Supplemental irrigation is required for maize, millet, and sorghum in most years, but it is not routinely available. For none of these crops are chemical fertilizers or mechanical traction available. Climate and rudimentary technology make for large swings in production. FAO data (as reported in Osborne, 1996) show that over the period 1979–94 total cereal production averaged 5,088,000 mt with a standard deviation of 1,040,000 mt.

In addition to the diminishing size of operational landholding per farm family, so too the per-family availability of livestock and draft animals is declining (Diriba, 1995). This has two consequences: the amount of traction power and of organic fuel and fertilizer per family is diminishing; and the availability of animals as a store of wealth that can be sold in times of production failures is declining.[5]

NO EASY WAYS OUT

In 1989 the Socialist Government of Ethiopia announced a National Food and Nutrition Strategy and a year later a national irrigation policy (Kloos, 1991:296; Abate, 1991). Its goal was food security in the conventional sense of national self-sufficiency. As noted, corollary goals were to double food production, establish food reserves against famine, and to improve average caloric intake dramatically. When one examines the means to achieve these goals, one has square pegs for round holes.

Most outside advisers from various parts of the donor community emphasize the complicated task of intensifying production in the highlands; that is, they recommend taking hold of the "traditional" systems of cultivation and making them more productive. To do so requires a great deal of adaptation to local conditions coupled with improvements of technology that the farming population can afford.[6] Here are some of the specific measures:

> *Disaster Prevention.* A minimalist definition of food security is whatever it takes to prevent famine and widespread death (World Bank, 1996). Some measures can be taken immediately; others can be begun immediately but their full effects will not be felt for some time. For example, improved grain storage, augmented local food processing, reduction of post-harvest crop losses, the constitution of national grain reserves, and planning the logistics of emergency relief are all steps that can be taken in the short term to avert disasters (Jansonius, 1989). Identifying the most at-risk fami-

lies (as well as localities) and providing coping strategies for them can be begun immediately, but results will be registered only in the middle term.

When one views food security as the avoidance of famine, then there are many initiatives to be taken that do not directly bear on agricultural production itself. Developing supplementary employment (the classic approach is food for work) and off-farm employment opportunities can provide the cash flows that allow families to acquire food in times of crises. But there are also policies that can induce changes in cultivator practices, such as changing or liberalizing prices in the agricultural sector by removing subsidies and allowing producer prices to rise to their market-determined level.

Water Management and Harvesting. Subsumed under this rubric are the construction of microdams to capture flood runoff for use in supplemental irrigation in the dry season and for watering livestock; the construction of bunds and check dams to slow and divert the devastating runoff of heavy rains in the *mehr;* the rehabilitation of watersheds through planting trees and grasses, and terracing fields.

In fact, both historically and in the past two decades, the northern highlands, especially in Gondar and Tigray, have witnessed substantial efforts along these lines (see Lee et al., 1996). Thousands of microdams are projected for construction using local labor and materials. Check dams and bunds have also been constructed in many areas of the highlands. The results have been at best mixed. Howard (1995) has calculated the extent of such projects for the period 1976–90 and their survival rates. For example, only 22 percent of new tree species survive, and only half of all afforestation survives. Bunds, check dams, and terraces in their majority are destroyed by floods and runoff. In addition, cattle left to graze on harvested fields break down the bunds and terraces. Constable and Belshaw (1989) argue that erosion is affected not as much by deforestation as it is by plowing during the short rains and by the grazing habits of livestock.

Microdams have not fared much better. They too are frequently swept away or eroded by the annual rains and floods. While intact, their reservoirs, as intended, attract animal populations that trample the groundcover around them, leading to more rapid sedimentation and to the contamination of the stored water. Because the construction and maintenance of all these innovations requires a continuous effort on the part of local populations, the collective action problems that arise may prove insurmountable. Farmers may want to reserve their labor for their own fields and animals even at the cost of losing the (hypothetical) benefit of a collectively provided good.

Improved Cultivator Practices. These would consist in familiar seed and fertilizer packages to take advantage of the improved water management and harvesting. Neither improved seeds nor chemical fertilizers are available to the bulk of Ethiopia's farmers now. The Global 2000 project, sponsored by the Sasakawa Foundation with the patronage of Jimmy Carter and Norman Borlaug, has been addressing this problem since 1992 in some of the better watered regions of Ethiopia with very positive results (see *Economist,* November 25, 1995, pp. 41–42). Can it work in regions of less reliable rainfall or anywhere that increased production cannot be readily marketed because of poor infrastructure?

Electricity, Mechanization, and Off-Farm Employment. One of the great successes of the People's Republic of China was in bringing electricity to China's villages and towns. This allowed both the mechanization of some agricultural tasks (such as pumping, milling, and threshing) and the creation of off-farm employment in agro-processing and light manufacturing.

Surely this should be a goal for Ethiopia, but, given the size and nature of the terrain and the poverty of the potential beneficiaries, costs will be very high and returns to investments problematic.

Move Population to the Lowlands. Ethiopia's population has always been concentrated in the highlands. Up to 1,500 meters above sea level, malaria, trypanosomiasis, and other disease vectors flourish. The highlanders have always regarded the lowlands as regions of heat and pestilence, equally lethal to draft animals and to humans (for graphic evidence see Cheesman, 1968).

Nonetheless, the western lowlands along the Sudanese border and draining the Blue Nile, Dinder, Rahad, Seteit, and Atbara rivers, comprise thousands of square kilometers of relatively flat alluvial plains that are ideal for mechanized agriculture and, as we shall in Chapter 5, irrigation. Even without irrigation, it has been hoped that these lowlands could provide quickly marketable surpluses of oil seeds, pulses, maize, and sorghum.[7] Moreover, after the 1983–85 famine, the *dergue* ordered the transfer of more than 1 million highlanders into the lowlands, although the chosen region was actually in the Akabo-Baro basin (see Ministry of Agriculture, 1987), to the south of the Blue Nile watershed. In short, food security, it was believed, could be attained through the movement of vulnerable populations to the lowlands to provide marketable agricultural products for the entire country.

As soon as the *dergue* collapsed, if not before, the forced migration of highlanders was abruptly reversed. Most preferred the crowded but familiar highlands to the heat, disease, and unwelcoming populations of the lowlands. A resurrection of private, mechanized farming in the lowlands, with or without irrigation, could generate surpluses of oil seeds, maize, and sorghum. Without a good road system, storage, and robust markets, such schemes may not make good economic sense.

> *Trade for Food.* A broad definition of food security would include a strategy for trading in international markets for whatever is not produced locally. In other words, rather than seeking self-sufficiency, Ethiopia should seek ways to generate foreign exchange adequate to meet its needs, especially in the event of a production collapse. This would mean providing for the 10 percent or so of total consumption that is currently imported in the form of food aid in average years, and for 20 percent (or upward of 2 million tons) in bad years.

This will be a long-term challenge. In the early 1990s, coffee exports accounted for over half the total value of Ethiopia's exports, and raw materials for another 20 percent. Nontraditional exports accounted for only 10 percent. This definition of food security would emphasize agricultural production for export in order to cover needs in food imports.

One implication of this reallocation of resources is that lowland agriculture would be destined mainly for export, not for provisioning the highlands in basic foods. Even if such a decision were taken—and the current Ethiopian government shows little inclination to move in this direction at the expense of self-sufficiency—the Uruguay Round and WTO strictures may cause the terms of agricultural trade to shift in favor of "temperate" products (wheat, corn) and against "tropical" products (coffee, tea, etc.; see Raffer, 1997:1902–3). Whether for export or for sale in the highlands, lowland production still faces all the infrastructural problems in storage and transport alluded to above.

PULLING THE PIECES TOGETHER

Ethiopia is likely to try to make progress on all fronts. There may be a strong temptation to outflank the difficult socioeconomic context of the highlands through large-scale irrigation schemes in the lowlands. But if avoiding another famine is the primary objective, the task of intensifying highland agricultural production cannot be avoided. For this endeavor to succeed, progress must be made on several interrelated fronts; sustainable intensification must come as a package of approaches or it will not come at all.

Water harvesting and management will be effective only if accompanied by rehabilitation of watersheds. That in turn entails managing herds and grazing, afforestation, bunding, and terracing. For some time it must be accepted that earthworks will erode and wash away. Ethiopia's abundant factor is cheap rural labor, which should be used for the repetitive tasks of watershed improvement and construction of earthworks. To so state does not, however, answer the question as to who organizes the labor and with what incentives. Cash rather than food for work is the preferred incentive, and the donor community can obtain a great deal for very little money by supporting rural works programs. At some point, local communities will have to see it as in their interest to maintain what has been constructed or planted.

The highlands are broken and fissured. Irrigation schemes must be designed for small perimeters and widely varying soils. Bottom lands in valleys and alluvial deposits offer the best opportunities but not economies of scale.[8] Those who benefit from microdams should be made to share in the cost of their construction. That said, there may be no more than 300,000 irrigable hectares in the highlands, or about 5 percent of the currently cultivated surface. That amount of area, no matter how predictable its production, cannot begin to meet incremental food needs. All of highland agriculture needs to be made more productive.

It appears particularly important to make available supplemental irrigation, through either surface delivery systems or tube wells, during the short rains. Predictable water supply at that time of year (March–April) can lengthen the growing season and allow greater cultivation of sorghum and maize.

Crop diversification and crop alternance, or intercropping with nitrogen-fixing plants, will reduce risks and protect soil nutrient levels. To be successful, these changes require a competent and motivated extension service, something that does not exist at present. Moreover, if new seed and fertilizer packages are introduced, along the lines of the Global 2000, competent extension will be all the more necessary. These packages, be it noted, will only work well if there is assured water supply.

Access to or ownership of livestock and draft animals is another requirement. Credit programs for animal purchase could meet part of that requirement. Yet success could have important negative effects in overgrazing and destruction of earthworks and terracing. The growth of animal herds must be integrated into local level strategies for protecting groundcover and water sources. Again, one is faced with difficult challenges in monitoring animal herds.

The central and regional governments must undertake the construction of storage facilities for marketed grain, both as a reserve against production failures and to smooth out market prices for grain. Storage in turn needs to be accompanied by the extension of the road system so that the geographic borders of markets can be expanded and grain moved from surplus to deficit regions.

Education must be provided to rural populations in anticipation of much more sophisticated agricultural practices within a generation, and to prepare rural youth to find work in the off-farm labor force.

The biggest policy issue facing the current government is whether or not to maintain public ownership of agricultural land. The current government has resisted pressures from the international donor community to privatize agricultural land. As Mexico discovered in its communal, or *ejido,* sector, it appears that in Ethiopia widespread, illegal sharecropping and subleasing is going on (Diriba, 1995:353). But even an informal private property regime cannot provide de facto owners collateral in credit markets nor incentives to invest in land to which they do not have de jure title. Under the circumstances, mining the land is still the most likely response (see Donovan, 1996).

Of nearly equal importance are three economic changes: to redesign the tax burden on the agricultural sector to encourage production, to reduce taxes on exported commodities, and to liberalize prices to augment the amounts of production that are marketed.

CONCLUSION

Ethiopia is still at the beginning of its structural transformation from an agriculture-based economy to one based on manufacturing and services. It does not have an abundance of natural resources nor a market-friendly geography to hasten the transformation. It does, potentially, have an abundance of usable surface water and attendant hydropower. It is tempting to see them both as the solution to increasing agricultural productivity, avoiding famine, and providing the energy needed for the non-farm and off-farm service and manufacturing sectors. Surely, over the long haul, say twenty-five years, this vision is right. To pursue it aggressively now would also send a powerful message downstream regarding Ethiopia's pressing needs and its insistence upon equitable use of the Nile. Yet in the short and medium term, big projects for power and irrigation will not address the food needs of Ethiopians directly. Only the immediate introduction of carefully designed strategies to increase highland agricultural production on a sustainable basis can act as a hedge against the next precipitous drop in rainfall.

The simple fact is that agriculture and agricultural production remain the backbone of the Ethiopian economy. Food security is a strategic and a political goal, and in all scenarios more intense use of the western Nile watershed is part of the strategy. Production failures in the agricultural sector have spelled doom for two political regimes in the past thirty years, a lesson Meles Zenawi appeared to have forgotten as he poured scarce national resources into a war with Eritrea. Had he not scored some major victories in the spring of 2000, he would have faced a bitter political harvest in the fall.

Chapter 5 The Imperfect
Logic of Big Projects

There are two strategies by which to enhance food security. They may be combined in varying proportions. The first, preferred by past Ethiopian governments and by most governments everywhere, is the big project, with its top-down, technocratic approach. Large water storage and power generation projects are believed to offer the quickest and most lasting solutions to increasing agricultural production, food processing, and off-farm, rural employment. The other strategy is to pursue small-scale, local initiatives in water conservation and harvesting, careful extension work with clusters of farmers, improved local storage, expansion of road systems, and the ability to move food to the regions where it is most needed. In this strategy some economies of scale are lost, but it is less capital intensive, involves local producers and extension agents, and may foster sustainable, community-based production.

Each strategy or combination thereof seeks to harness Ethiopia's existing surface water resources. The statistics of Ethiopia's water balance are impressive. It is estimated that, on average, some 1.3 trillion cubic meters of rain fall annually over its extensive surface. This amount generates about 112 billion cubic meters (bcm) in surface

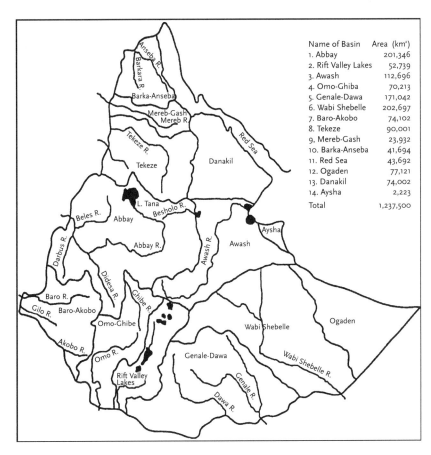

Name of Basin	Area (km²)
1. Abbay	201,346
2. Rift Valley Lakes	52,739
3. Awash	112,696
4. Omo-Ghiba	70,213
5. Genale-Dawa	171,042
6. Wabi Shebelle	202,697
7. Baro-Akobo	74,102
8. Tekeze	90,001
9. Mereb-Gash	23,932
10. Barka-Anseba	41,694
11. Red Sea	43,692
12. Ogaden	77,121
13. Danakil	74,002
14. Aysha	2,223
Total	1,237,500

Figure 5.1. Ethiopia's River Basins

Note: This map does not show Eritrea as an independent country.

Source: Zewdie Abate, *Water Resources Development in Ethiopia: An Evaluation of Present Experience and Future Planning Concepts* (Reading: Ithaca Press, 1994), p. 29.

runoff, of which half is technically "storable." However, according to one's reckoning, 9 to 11 rivers annually discharge across Ethiopia's borders some 100 bcm to its neighbors (Figure 5.1).[1] Theoretically, Ethiopia holds extraordinary potential for large-scale, irrigated agriculture and for hydropower generation. In point of fact, only a tiny fraction of this potential has been developed.

The gross statistics on supply would thus seem to yield three conclusions. Part of the solution to Ethiopia's recurrent famines must lie in extensive irrigation and predictable agricultural production. Second, if the country is not to be chained to its agricultural base, its water resources, in the absence of petroleum or coal reserves, must be used to produce hydropower for industrialization. Finally, if the first two assumptions hold, then Ethiopia's neighbors will have to expect, *ceteris paribus,* a downward shift in surface water supply. As we have seen in Chapter 4, micro projects can, over the very long haul, deal with some of the challenges, but, as noted, a preponderant body of opinion within Ethiopian policy groups argues that big projects enjoy economies of scale and implementation schedules that represent the most rapid and effective response to the country's structural problems.

The logic is powerful, if not overwhelming, but it is seriously flawed. If Ethiopia's only problems were inadequate seasonal water supply and large, interyear variations in rainfall, then large-scale water storage projects would make eminent sense. The real problem is that optimal storage sites, from the point of view of irrigation *and* power generation, are not proximate to existing markets, reliable road grids, or ports. These same sites, moreover, are the most prone to rapid sedimentation and to relatively high rates of surface evaporation. In short, the economic rates of return on such projects are uncertain, the engineering challenges formidable, and the lifetimes of the projects themselves potentially short.

Ethiopia would do well to reflect on the history of the Sudan since about 1930. There, under much more favorable circumstances from the point of view of topography, domestic and international markets, infrastructure, and ports, first the British and then the independent Sudan initiated and expanded what the Sudanese are proud to call the largest irrigated farming scheme under single management in the world. First as the Gezira scheme, and now as the Gezira-Managil scheme, more than 2 million acres were brought under year-round irrigation in the inverted pyramid lying between the Blue and White Niles, and with its apex lying at Khartoum. To this giant irrigated perimeter the Sudanese have added in the 1960s and the 1970s the Khashm al-Girba project on the Atbara and the Kenana sugar perimeter on the White Nile. These schemes have not insulated the Sudan against periodic regional famines, as witnessed in the early 1980s, nor have they provided a firm anchor in international markets for cotton and sugar. Their hydropower capacity is not sufficient to supply the cities of Khartoum-Omdurman, and that capacity has been dwindling with the rapid sedimentation of the dams at Roseires and Khashm al-Girba.

THE CURRENT SITUATION

In the early 1990s it was estimated that Ethiopia possesses about 3.7 million hectares (9.2 million acres) of irrigable land. This figure probably includes Eritrea's irrigable surface, which is not extensive. At that time, about 97,000 hectares were irrigated by modern techniques, and of those 71,000 were located in Ethiopia's pioneer scheme in the Awash Valley (see below). Another 68,000 were under development, while 60,000 were irrigated by traditional methods. By whatever method, less than 3 percent of Ethiopia's potential was actually under irrigation (Z. Abate, 1994; EVDSA, 1992).

The story is the same with respect to hydropower generation. Only about 2 percent of Ethiopia's hydropower potential is currently exploited (Federal Democratic Republic of Ethiopia, 1996). Total installed capacity, both thermal and hydroelectric, is 400 MW, of which 338 MW are part of the national grid. About one-quarter of this is generated at three dams in the old Awash project, while the more recent dams at Finchaa and Melka Wakane generate another two-thirds. Within the next twenty-five years, Ethiopia hopes to quintuple its hydropower generation to 2,200 MW (Ministry of Water Resources, 1996). The most promising sites are on the Omo, the Akabo-Baro, and the Abbay and its main tributaries.

The EVDSA-WAPCOS study of 1990 put Ethiopia's total undeveloped hydropower potential at 85,000 GWh per year.[2] A little more than 51,000 GWh could be generated in the Abbay (Blue Nile basin), although the Akabo-Baro and the Omo-Ghibe, while capable of producing less power, can do it more reliably than the Abbay because of less variation in flow (Ministry of Mines and Energy, 1986: 50). Studies conducted by a number of consultants in the 1980s, including WAPCOS (1990), Ansaldo-Finnmecanica (1986), and Acres International for the EELPA (1982), all concurred that the potential of the Teccaze was and presumably still is inferior to all other basins in the western watershed. This is due to the high interyear variability of flow, consistently high seasonality of flow, and steep gorges that accentuate sedimentation problems. In contrast, the Tecazze drains Ethiopia's most drought- and famine-prone region, and one that sheltered the successful guerrilla movement that led to the overthrow of Menguistu Haile Meriam and the *dergue*. One should not be surprised, therefore, that the first major storage site to be developed by the new Federal Republic of Ethiopia lies on the Teccaze in the Tigray region.

All the most promising hydropower sites, and even those in the upper Teccaze, are distant from the major sources of demand, around Addis Ababa and Dire Dawa. In the studies done throughout the 1960s, 1970s, and 1980s, sites on the Teccaze were examined for their potential to supply power to Asmara in Eritrea. At that time Eritrea was a (highly contested) province within Ethiopia. Today it is an independent country. That fact does not mean that power development on the Teccaze could not supply Eritrea, which, for a time, was a very friendly neighbor, but Eritrean independence has changed the parameters of any such projects and complicated the issue of establishing markets and payments for power delivered. Similarly, hydropower sites on the Abbay and the Akabo-Baro could be seen in two lights. They could be projects by which supply of power elicits demand in the hitherto remote, off-grid parts of Ethiopia with unreliable agricultural bases. But, more important, they could supply the great power deficit areas of neighboring Sudan in the Khartoum-Omdurman conurbation and Wad Medani in the heart of the Gezira scheme.[3] The basic point is that Ethiopia has not yet seen itself as a major exporter of hydropower. Consequently the cost-benefit analyses of potential projects have been carried out on the basis of domestic demand for power and irrigation or potable water.

SEDIMENTATION

It is a commonplace that Egypt's agricultural lands and soils ultimately originated in the Ethiopian highlands, deposited by the annual floods over hundreds of thousands of years. The observation tells us that in semi-arid areas of the world, watersheds are particularly vulnerable to erosion and heavy sediment loads. The problem is exacerbated by the fragility of groundcover. In the past century or so, the increasingly intense use of steep slopes for cultivation and grazing, combined with the search for firewood and other combustible materials by growing populations, have accelerated the rate of erosion just about everywhere in the so-called semi-arid tropics. The Achilles heel of large-scale water storage projects in these regions is sedimentation.

When cost-benefit (and of course, design) analysis is run on a proposed dam and reservoir, crucial assumptions are made about sediment load. The reservoir is conceived of as having a certain area for "dead" storage to receive sediment deposits over time. Above the dead storage area is the live storage area, that is, the water that can be used for irrigation and power generation. Once the dead storage area has been filled with silt, the live storage area diminishes, year by year, and the effective life of the dam and reservoir approaches its end. It is therefore crucial in estimating the returns to an investment in such projects to have a reasonable estimate of the rate at which sediment is likely to be deposited. In reading many such estimates over many years with respect to projects in the Nile basin, I have found that the problem is systematically minimized or avoided altogether (with calls for further study). Many important interests want large-scale projects to go forward: host governments, to show that they are working to better the lives of their people (and if there are percentages on large contracts to be spread around officialdom, so much the better), suppliers of equipment such as turbines and transmission lines, bilateral and multilateral donors concerned with promoting exports of goods and engineering services and/or with showing *their* commitment to the well-being of a given country's population, and, of course, all of the private and sometimes public firms that will obtain some piece of the construction action. None of these interests wants a project to be blocked on cost-benefit grounds because an overly prudent estimate of sedimentation rates lessens the useful life of the project.[4]

There is some, misguided, belief that even if a site fills with silt more rapidly than expected, one still gains agricultural land in the form of the deposited silt. But, as Goldsmith and Hildyard (1984:230) have argued, silt deposits under the heavy weight of stored water compact and form hard pans. The only cultivable areas tend to be in midstream, where coarser, less compacted soils are deposited. Moreover, most of the best sites for dams on major rivers are known, if not already developed. Abandoning "silted-up" sites and moving to others is not really feasible.

With respect to the western watershed of Ethiopia, there is little doubt that sedimentation could pose a serious threat to the viability of many proposed projects. For graphic evidence of what might occur, one has only to look further downstream on the Blue Nile (Roseires Dam) and the Atbara (Khashm al-Girba Dam) in the Sudan. The two dams just cited have silted up at rates far exceeding the estimates made at the time of construction. Roseires, which was designed to provide irrigation water for the Gezira-Managil scheme and power for Khartoum-Omdurman, is no longer able to perform either task adequately. In planning for Roseires, designers estimated sedimentation at 15 mcm per year; in the reservoir's first ten years of operation, the real rate was 55 mcm per year. By 1985 nearly all the dead storage capacity of the Roseires reservoir had filled with sediment (Badawi et al., 1997:8).

In Ethiopia's eastern watershed, the series of dams in the pioneering Awash project area, but especially the Koka Dam, have experienced rapid rates of sedimentation.[5] The only serious estimates I have come across of sedimentation rates in the western watershed were those prepared by the Bureau of Reclamation study in the early 1960s (see U.S. Department of the Interior, 1964:appendix III, "Hydrology," pp. 11–20, and vol. 2, "Plans and Estimates," pp. 351–404). This study provides data site-by-site, and estimates of live and dead storage are made on the basis of measured sedimentation loads in different sub-basins during periods of flood (when loads are highest) and low water. On average, suspended matter may be as much as 2 percent in weight of the total water discharge. For argument's sake, let us return to the figure of, on average, 100 bcm flowing across Ethiopia's borders annually. Let us further assume that of that amount 70 bcm is in the western watershed. That would mean that on average about 1.4 billion metric tons of silt is transiting downstream annually to the Sudan. Any projects in the western watershed will capture some part of this load. The question is, for any specific site, how much and how quickly?

Sometimes the regime by which the reservoir is operated may alleviate the problem. For instance, the reservoir may be used for seasonal storage only. When the crest of the flood approaches, the reservoir is partially or totally emptied, the flood is allowed to flush the reservoir and to pass through, carrying with it the bulk of the annual load of sediment and some of the stored sediment. The sluice gates are then shut to capture the tail of the flood, with less sediment load, and to refill the reservoir. This is how Roseires and Khashm al-Girba are operated, but they both have nonetheless suffered from severe siltation problems. Such storage practices, moreover, leave one vulnerable to successive low years during which the capture of the tail end of the flood is not adequate to refill the reservoirs. The Aswan High Dam in Egypt was built precisely to overcome the seasonal storage problem and to provide a reservoir of such capacity that successive floods could be stored in their entirety *and* dead storage provided for two hundred years of sediment deposits.

The sites least vulnerable to rapid sedimentation in Ethiopia's watersheds are those at the highest elevations and with the least slope in the watercourse. These sites, however, are furthest from lands suitable for extensive irrigation and, because of the gentleness of the watercourse slope, not optimal for hydropower generation.

SURFACE EVAPORATION

Only marginally less significant than sedimentation is the problem of surface evaporation at reservoir sites, and of storage losses in a more general sense. Losses are determined by the geological characteristics of reservoir locations, air temperatures, wind velocity, and, of course, average precipitation at the site. As a general rule, sites in areas with prevailing high temperatures, high winds, and low precipitation will experience the highest rates of surface evaporation. The problem may be exacerbated if surface-to-volume ratios are high; that is, if the storage site reservoir is wide and shallow, increasing the extent of exposed surface to the amount of water stored. (It is better to store water in a glass than on a plate.) Surface losses may also be compounded by aquatic weeds and grasses. This has been a problem in some of Ethiopia's existing reservoirs. In other locations in the Nile system, lake surfaces and irrigation canals have become choked with water hyacinth, the growth of which may double the rate of surface evaporation (see Chapter 7).

There is no doubt that surface evaporation in the Ethiopian highlands is less than it is in the border areas with the Sudan, and less than is the case at the reservoir at the Aswan High Dam. Thus, storing water at higher altitudes will yield some savings in water, but the savings may be somewhat overstated (see Guariso and Whittington, 1987). Throughout the western watershed and at all altitudes, surface evaporation is high. Even taking into account average rainfall in the Ethiopian highlands, at any given storage site there may be a net loss of water, compensated only by surface runoff into the reservoir. Let us take Lake Tana, long coveted by Great Britain as a natural storage site by which to regulate the flow of the Blue Nile.[6] Rainfall over Lake Tana ranges on average between 1.0 and 1.5 meters per year, while surface evaporation ranges between 1.2 and 1.3 meters. There is little net gain to Lake Tana from rainfall itself. Maintaining the lake level depends upon surface runoff. The Teccaze watershed is in chronic water deficit. Average surface evaporation is 2.0 meters while rainfall averages 1.0 to 1.5 meters. The Abbay basin shows a range of surface evaporation rates from 1.3 meters in the highlands to 2 meters at lower altitudes. Only in the southwestern Akabo-Baro basin is there a significant net water surplus: rainfall averages 2.2 meters, while surface evaporation averages 1.0 meter in the highlands and 1.7 meters in the lowlands and swamps (Ministry of Mines and Energy, 1986:43–47). Woundoneh (1996) gives somewhat higher average surface evaporation rates (Table 5.1).

When one combines the high seasonal variability of flow, the high rates of sedimentation, and the relatively high rates of surface evaporation, the triple challenge can be met only by large storage sites with multiyear storage capacity. Only the southwest and the Omo basin (not part of the Nile watershed) escape this general rule. In the two latter basins, small, run-of-the-river hydropower projects may be feasible, but all potential sites are too far from centers of power consumption to make the projects economically attractive. As for the Teccaze, where seasonal flow may vary from 4 to 1,000 cubic meters per second, the possibilities for development are "very slim indeed" (Ministry of Mines and Energy, 1986:43, 47; EELPA, 1982:110).

Reservoirs situated in areas of permeable soils and rock strata will lose water

Table 5.1 Annual Rainfall and Evaporation at Various Altitudes in Ethiopia and at the Aswan High Dam (in Millimeters)

Altitude	Abbay		Teccaze		Akabo-Baro		Aswan High Dam	
Meters asl	Rainfall	Evaporation	Rainfall	Evaporation	Rainfall	Evaporation	Rainfall	Evaporation
2,200	1,608	1,140	552	1,356	2,316	1,068		
1,700	1,116	1,404	636	1,560	2,292	1,140		
600	900	2,700	612	1,764	1,212	1,680		
160							0	2,683

Sources: The figures for rainfall and evaporation for 600-meter altitude in the Abbay basin are from Woudeneh Tefera, "Water Demand Scenario in the Ethiopian Portion of the Nile Basin," Fifth Nile 2002 Conference, Addis Ababa, February 24– 28, 1997, p. 5. His figures are on average higher than Ministry of Mines and Energy (1986, Supplementary Report 3:215–50) which is the source for all the remaining figures for Ethiopia. The comparative evaporation figures for the Aswan High Dam Reservoir are drawn from Mahmoud Abu Zeid, "Environmental Impacts of the Aswan High Dam," *International Journal of Water Resources Development,* 5, no. 3 (September 1989): p. 152.

through seepage. Undetected fissures opened by the pressure of the stored water will increase such losses. Water stored in active earthquake zones, as in parts of California, may be exposed to sudden and catastrophic loss. Any storage project must be based on reasonable estimates of likely storage losses, but, as in the case of sedimentation, there may be a systematic bias toward underestimating the problem.

The best storage sites in Ethiopia from the point of view of minimal sedimentation, surface evaporation, and seepage are in the highlands. Several million years ago, when Ethiopia and the Rift Valley were the locus of massive volcanic activity, the highlands were covered in an impermeable basaltic cap. The basalt, in turn, affords storage sites that suffer little from seepage and watercourses that carry relatively little sediment load. It is only when rivercourses dip below 2,000 meters that the basalt gives way to more easily eroded rock and soils. As rivers descend toward sea level, temperatures and aridity increase, along with sediment loads. It is also at or below 1,500 meters above sea level that malaria becomes endemic. In the southwest lowlands, malaria is accompanied by tsetse fly and sleeping sickness. Humans, cattle, and draft animals all suffer.

Ethiopia's conundrum is that water stored in much of the highlands cannot be economically used in irrigation (given present technologies) over the rough and broken terrain typical of higher elevations. And, because the slopes of watercourses in the highlands are generally gentle, the hydropower potential is less than at lower elevations. By contrast, the best irrigable lands lie at low altitudes on the floodplains along the border with the Sudan where temperatures, wind velocity, and sediment loads are particularly high. We need to explore this conundrum in some detail.

THE AWASH VALLEY: THE PROTOTYPE

My focus in this chapter is upon Ethiopia's western watershed, but the simple fact is that in terms both of history and of twentieth-century development, Ethiopia's face has always been turned eastward. This is reflected in one of its oldest urban centers, Harar, sitting atop the eastern escarpment of the Rift Valley and dominating all trade moving up from the Red Sea and the Indian Ocean. This axis was strengthened in the early part of the twentieth century with the construction, by the French, of the rail line from Djibouti (which became Ethiopia's major point of access to the sea) to Addis Ababa, by way of Dire Dawa lying close to Harar. It was as governor of the Harar and the eastern region that Ras Tefari, who was to become Emperor Haile Selassie, learned the political skills and built the alliances that allowed him to ascend to the throne.

The first steps toward developing the Awash Valley, which is part of the Rift fault line, occurred during the Italian occupation. The Awash River flows from southwest to northeast, petering out in Lake Abe on today's border with Djibouti. It never reaches the sea.

In the 1950s the Kingdom of Ethiopia, solidly aligned with the United States in the Cold War, launched an integrated rural development scheme, modeled, as was common in many developing countries at that time, on the Tennessee Valley Authority (see Chapter 2). The complex of dams, plantations, and agro-industries established in the Awash valley was put under the AVA—the Awash Valley Authority (Walker, 1974).

The object in the Awash valley was to construct multipurpose dams to generate electricity for Addis Ababa and Dire Dawa, irrigate cash crops (primarily cotton and sugar cane), develop roads and social infrastructure, and settle the Afar nomads. Over the period from 1960 to 1971, three dams were built and commissioned: Koka and Awash II and III. The first dam was financed from $16 million in reparations paid to Ethiopia by Italy. Italian firms were awarded the construction contracts. The reservoir storage capacity at Koka is about 1.6 bcm.

The installed generating capacity of the three dams is 440 GWh per year, and eventually a little over 70,000 hectares of land was brought under irrigation. As already noted, Koka in particular, as the upstream dam in the system, has experienced fairly severe rates of sedimentation. It also has been afflicted by an invasion of water hyacinth (the hyacinth "scourge" will be discussed in greater detail with respect to Lake Victoria and Uganda's Owen Falls Dam). The Awash system has, as a result of reduced capacity, suffered from occasional irregularities in power production. Two more recent projects, the Finchaa-Amarti in the Abbay (Blue Nile), and the Melka Wakane in the southeastern watershed, completed in 1973 and 1989, respectively, have a capacity of 1,170 Gwh/year between them, or well over twice the capacity of the three Awash power plants.

THE BUREAU OF RECLAMATION BASELINE

Between 1958 and 1963, the Bureau of Reclamation of the U.S. Department of the Interior, carried out an extensive survey of Ethiopia's Blue Nile watershed. The study was conducted on behalf of the Ethiopian government, and its timing and implications were relevant to the Cold War games being played in the Nile basin (see Chapter 3).[7] The survey was published in seventeen volumes in 1964. It helped train an entire generation of Ethiopian hydrologists, and it gave rise to the Ethiopian Water Resources Department, which later became the Ethiopian Valleys Development Studies Authority.

The survey identified twenty-six storage sites along or near the 900 kilometers of the Abbay from Lake Tana to the Sudanese border. Several of the sites were on the major tributaries feeding into the Abbay. Most of the projects identified were multipurpose, combining hydropower generation with irrigation. Potential power generation was estimated to be 38 billion kWh or 8.6 million kW at 0.5-plant factor (U.S. Department of the Interior, 1964, vol. I: 97). An area of 438,000 hectares of newly irrigated land was projected, and the total reduction in Abbay flow into the Sudan was estimated to be 5.4 bcm (or a little more than 6 percent of the total average annual discharge of the Nile River). Some of the projects did not involve power generation: the Megeche scheme, northeast of Lake Tana, was to provide pump irrigation on 25,000 hectares. The Megeche area is one of the relatively flat sites in the highlands that is conducive to surface irrigation. It can be contrasted with the site in the deep gorge and rough terrain at Karadobi on the Abbay itself.

One of the oldest and most ambitious projects, mentioned by Major Cheesman in the 1930s as having already been studied (Cheesman, 1968:221), is the Tana-Beles. There is a point on the southwestern shore of Lake Tana where an escarpment only five km wide divides Lake Tana from the basin of the Beles River, which then feeds into the Abbay not far from the Sudanese border. A tunnel through the escarpment would drop water from Lake Tana into the Beles. The drop would be exploited to generate power, and some 65,000 hectares could be irrigated in various parts of the Beles watershed. The Bureau of Reclamation identified one other site on the Beles for power development.

This site has been studied subsequently. The 1982 EELPA survey designated it, along with the Lower Didessa site, as one of the most promising in the highlands. In 1986 a regulator at the outlet of Lake Tana was completed. It would have enabled EELPA to maintain a lake level sufficient to feed the diversion channel and tunnel to the Beles basin. The EVDSA-WAPCOS study of 1990, however, ranked this project (by that time including five storage sites along the Beles itself) as fourteenth out of fifteen Abbay projects. Despite the ranking, the Menguistu regime nonetheless contracted with the Celini Company of Italy to undertake the project. By that time, it was decided to irrigate 150,000 hectares with water stored along the Beles. For reasons I have not been able to determine, the project was abandoned, leaving behind only an airstrip constructed by the Italian contractor.

The Abbay's largest tributary (25 percent of the total discharge) is the Didessa, on which the Bureau projected four reservoirs. Three were to be multipurpose, while the furthest downstream was to be for power only.

The Dinder and Rahad tributaries, which join the Blue Nile only after crossing the border into the Sudan, were to be the sites for multipurpose schemes that would irrigate extensive areas of the flat, alluvial plains along the border. We shall return to this and other lowland irrigation schemes below. The largest power project was the Karadobi, upstream of the confluence of the Guder and the Abbay, and, if completed, capable of regulating the flow of the Abbay. This site has, potentially, a highly favorable surface-to-volume ratio, located in a deep gorge with steeply pitched sides (see Figure 5.2).

The largest in storage volume of the projects identified by the Bureau was the so-called Border Dam, some 21 kilometers upstream from the Sudan. At this point the Abbay has dropped from 1,786 meters asl to 577. The proposed dam was and is in the extremely hot and arid lowlands where surface evaporation is at its maximum. The Border Dam was to have a storage capacity of more than 11 bcm (or nearly seven times the size of the Koka reservoir) and power generating capacity of 6.2 million kWh (or two-thirds of the actual power capacity of the Aswan High Dam).

With the exception of the Finchaa-Amarti project, none of the twenty-six sites identified by the Bureau has yet to be developed. Political turmoil and feeble state finances are the two principal reasons, but Ethiopia may be fortunate that it was unable to plunge ahead with the intensive exploitation of the western watershed. In light of what is now known about large dams and reservoirs, it is possible that several of the projects identified nearly forty years ago would make neither economic nor technical sense. The Bureau of Reclamation survey has, however, continued to serve as the baseline for all subsequent studies of that watershed, and it is a remarkable technical document.

We turn now to consideration of the Finchaa-Amarti, for it tells us more than any other project about the future of highland, medium-scale irrigation and power generation.

THE FINCHAA-AMARTI
MULTI-PURPOSE PROJECT

The Finchaa reservoir is located 170 kilometers northwest of Addis Ababa. The original site consisted of swamps and an impressive waterfall on the Finchaa River, a tributary of the Blue Nile. The Bureau of Reclamation identified it as a multipurpose project, but until a few years ago only the hydropower portion had been developed. Because the site is relatively close to Addis Ababa, it has become a principal source of the capital's electricity.

The project was begun in 1968, the corner stone laid in November 1970 by Emperor Hailie Selassie, and commissioning took place in 1972 (World Bank, 1969; EELPA, Nov. 1973; interviews with EELPA officials, spring 1996). The total value of the project was put at $34 million (in 1968 dollars) of which $23 million was in foreign exchange. The World Bank put up most of the foreign exchange, while USAID contributed $2.8 million. Harza Engineering conducted the first feasibility study, but no other U.S. contractors were awarded any part of the project. Typical of projects of that era, no environmental impact study was carried out.

If all highland sites could be as well suited to storage as Finchaa, then the future would be bright. The reservoir is about 2,200 meters asl so that surface evaporation is minimized. This is doubly important because the water stored in the first phase is a modest 900 mcm while the surface of the reservoir is relatively extensive at 175 square kilometers. There has been only moderate sedimentation at the site, but floating islands of weeds, set loose when the marshes were submerged, have proved to be a serious nuisance. A 3-kilometer tunnel beneath the dam channels water downhill to a surge chamber and to a 2.1-meter diameter penstock (or welded pipe) that takes the water downhill a further 1,310 meters to three turbines with 90 MW installed capacity.

From the outset, problems developed in the anchoring of the penstock. It is likely that during the rainy season there is heavy seepage of water into the excavated overburden on which the penstock rests. This was noted in 1976 by the first World Bank appraisal. It is still a problem today (interviews in EELPA operations division, April 1996). The penstock has lifted off its concrete anchor. EELPA is faced with a difficult choice. Either it must shut down the penstock, hence cutting off power to Addis Ababa, in order to reset the anchor, or construct another penstock and perhaps a second power station to accommodate planned expansion into phases 2 and 3 of the original project.

The initial step in phase 2 has already been taken, also in accord with the original Bureau of Reclamation study. In 1985 the Amarti River was dammed and diverted into the Finchaa so that something like 1.2 bcm are currently stored in the Finchaa reservoir. A third river, the Neshe, can be diverted into the Amarti to complete phase 2 and 3 of the water harvesting projects. Then the challenge will be to increase power generation at the site either by adding a fourth turbine at the existing power station or by building a new penstock and power station.[8]

In the meantime, the agricultural component of Finchaa is being added. In two phases, irrigated area for sugar cane will rise to 6,000 then 8,000 hectares. F. C. Schaffer and Associates of Baton Rouge, Louisiana, is constructing a refinery capable of processing 6,000 tons of cane per day. It will operate 250 days each year, closing during the rainy season when cane cannot be harvested. Once fully under way the plant will produce 80,000 tons of refined sugar per annum. The government will be responsible for operating the plant and for managing the sugar estate (i.e., the Awash model is being replicated). However, there is an interesting innovation. Sprinkler (pressurized) irrigation will be used rather than surface delivery canals. If this technique proves economically successful, it is far better adapted to the uneven terrain of the highlands than are surface (canal) delivery systems.

Not many sites in the highlands are as favorable to power generation and irrigation as are the Finchaa-Amarti-Neshe basins. There are, however, *some* other favorable sites, mainly in the rest of the Blue Nile watershed and in the Akabo-Baro basin. It is therefore somewhat surprising to find that the next major project in the western watershed will be carried out on the Teccaze, nearly universally seen as the least promising basin in that watershed.

TECCAZE 5

The Teccaze, over most of its course toward the Sudanese border, flows through steeply pitched gorges that facilitate storage (good surface-to-volume ratios) but are unsuitable for surface irrigation schemes. Moreover, the river is the most variable in annual discharge and seasonal flow of any in the western watershed. Finally, its most proximate source of demand for electric power is in Asmara, now the capital of an independent country. Nonetheless the Ethiopian Federation, under Meles Zinawi, has decided to implement a major hydroelectric scheme on the Teccaze. It is known as Teccaze 5, that is, one of five potential sites on the river for storage and power generation. It lies west-southwest of Mekele, the capital of Tigray province, where the river skirts the massive Simien mountain range that separates the Tigreña from the Amhara of Gondar and Lake Tana (Figure 5.2).

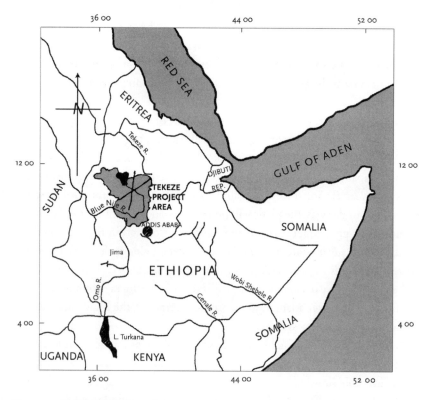

Figure 5.2. The Teccaze Project Area
Source: Howard Humphries, Coyne and Bellier, Rust Kennedy and Donkin; unpublished engineering consultants' report, draft, 1996.

The project entails a reservoir with 4 bcm of live storage and 4 bcm of dead storage. Estimated sedimentation rates would yield a thirty-six-year lifetime for the reservoir before dead storage began to eat into live storage. Initially, the power station will have installed capacity of only 72 MW, and Mekele will presumably be the major market for it (see Natural Resources Group, 1995; Ministry of Water Resources, 1996). As was the case with Finchaa some thirty years ago, the Ethiopian government has not sought agreement to this project by the downstream riparians. In the case of Finchaa, Egypt protested in vain. It has yet to be seen how Egypt will react to Teccaze 5.[9]

Because of the heavy fighting along the Ethio-Eritrean border between 1998 and 2000, work on Teccaze 5 has not progressed. Whatever the international ramifications, it has to be assumed that this project was endorsed by the Ethiopian government largely with the objective of stimulating economic life in Tigray in mind. On its own, this project would probably not pass muster. In terms of returns to investment to Ethiopia as a whole, there are substantially more favorable sites that remain undeveloped.

THE LURE OF THE LOWLANDS

Since 1900 the increasing commercial potential of lowland agriculture has drawn both private and public "entrepreneurs" from the highlands down to the baking plains along Ethiopia's long border with the Sudan. The lowlands were, and to some extent still are, infested with malaria and sleeping sickness. Native populations, such as the Hadendowa and Beni Amer, used the plains for grazing and subsistence agriculture.

The influx of highlanders mimicked an eastward movement of Sudanese commercial farmers from Khartoum and Wad Medani. The rail line linking Wad Medani to Port Sudan, and the much shorter line running from Mitsawa on the Red Sea through Asmara to Akordat, afforded a means to transport produce to cities and ports. The primary commercial crops of these areas were and are cotton, sesame, and sorghum. Once the tractor became available (draft animals had difficulty surviving in the lowlands), a pattern of using migrant labor, including, in the early years, slaves, and throughout, *fellata* of "Nigerian" origin, absentee investor-leasers, and local overseers developed on both sides of the border (McCann, 1990). In 1972 Mesfin Wolde-Meriam summarized the magnetism of the border areas in these words: "A very recent phenomenon is the rush to Setit-Humera, in the Angherib lowlands of Beghemdir near the Sudanese border. This alluvial plain, which until recently attracted no attention [cf. McCann, 1990, who shows the contrary], has suddenly become something like a new gold mine. Now more than 150,000 hectares are cultivated with more than 400 tractors, mostly producing sorghum, sesame and cotton. Very large numbers of labourers are flocking from Eritrea and Tigre, but also from Wello and parts of Beghemdir. The region that only about five years ago was nearly uninhabited is now teeming with hundreds of thousands" (Wolde-Meriam, 1972:116).

Under the *dergue,* which came to power a few years after Wolde-Meriam wrote, all the lowland areas were seen as prime sites for relocating famine-prone, highland populations on large state farms. None of the schemes survived, much less prospered. Today the new Ethiopian republic sees the same areas as logical zones for settling tens of thousands of demobilized soldiers from the old *dergue* forces, civilianized guerrilla fighters, and the large numbers of war and famine refugees from Eritrea, the Sudan, and western Ethiopia that have sloshed about the region for the past thirty years or more. Eritrea, for example, is coping with a return flow of refugees from the Sudan, estimated at nearly half a million, in addition to demobilized guerrilla fighters, and aims to settle them in the lower Gash plain. There is already an older irrigated cotton scheme at Alagadir which, with further water storage upstream, could be expanded.

Although fed by the alluvia of different rivers, the entire border area is characterized by broad, unbroken plains of vertisols (sometimes referred to as cotton-cracking soils). This expanse is hundreds of kilometers long, bounded in the east by the abrupt upthrust of the highlands and to the west only by the main Nile channel itself. There is one break in it, constituted by the high escarpment lying south of the Blue Nile drainage area and separating it from the Akabo-Baro drainage basin. We will consider the latter region separately because, although the issue is still one of lowland agriculture, the Akabo-Baro lowlands differ significantly from the northern border area. The 1964 Bureau of Reclamation study identified 230,000 irrigable hectares lying to the north of the Dinder and up to the Teccaze. A further 300,000 have subsequently been identified in the Teccaze-Anghereb catchments (i.e., the Seteit-Humera region to which Wolde-Meriam alludes; Natural Resources Group, 1995:60). Finally, the TAMS-ULG master plan study of the Akabo-Baro estimated an irrigable surface area of 335,000 hectares.

While these are hardly exact estimates, they are not wild. The combined surface area of the three zones is about 900,000 hectares or 2.25 million acres. That is the equivalent of one-third of Egypt's total cultivated surface as of 1996, and the fact that these acres are in the hot, relatively arid lowlands means that evapo-transpiration and crop water duties are far higher than in the highlands and not significantly lower than in Egypt or the Sudan. For the two northern zones stretching from the Blue Nile to the Teccaze, the gross water duty for irrigation of 530,000 hectares would be about 6.5 bcm per year, or well over 10 percent of Egypt's share under the 1959 agreement.[10] I will not factor in the 3 bcm that might be required for irrigation in the Akabo-Baro lowlands, as that is water that otherwise might have been lost anyway in the swamps downstream and across the Sudanese border.

From the point of view of cultivation, these zones present a number of advantages. The land is gently sloped, naturally smooth due to the layered deposits of alluvia over the millennia, and covered today, if covered at all, in scrub and savannah vegetation. The land can be easily cleared, surface irrigation systems readily installed, and gravity used to deliver water to the fields. The soils themselves are a mixed blessing. These vertisols are relatively rich, having seldom been farmed, they are very deep, and they retain water well. In contrast, when dry they crack deeply, so deeply in fact that Col. Cheesman wrote about his pack donkeys breaking legs in the cracks. When wet, the soils become viscous. It is impossible for man, beast, or machinery to operate on wet vertisols. The sandy particles in the soils can wreak havoc with plows and harrows no matter how hard the steel, a fact discovered by Dutch agriculturalists carrying out experimental farming of the vertisols in the Jonglei Canal region of the southern Sudan (see Chapter 6). The intrepid explorer of the Nile Samuel Baker described vertisols in 1861 as "pie crust" when dry and "pudding" when wet (as cited in McCann, 1990).

These soils have been cultivated successfully in the Gezira-Managil scheme in the Sudan but under conditions of relatively low rainfall and highly controlled surface irrigation. Where rainfall is higher and spread over a longer period, as is the case in the border lowlands, the tending of crops in the wet summer months must be done mainly by hand. It may be possible to prepare fields mechanically immediately after the first rains, but mechanized harvesting is still problematic.

The development of the lower Baro-Akabo presents very different challenges.

THE BARO-AKABO BASIN

The Baro-Akabo Basin drains an area of some 74,000 square kilometers. A high escarpment projecting westward from the Ethiopian highlands separates this basin from the Blue Nile and Teccaze basins further north. The rivers flowing southwest from the escarpment—principally the Baro, Akabo, Gilo, and Alwero—drain into the great Machar swamps of the southern Sudan and form part of the White Nile catchment.

Nearly all engineering and agronomic studies identify this basin as one of unusual promise. It has a variegated ecology with three distinct zones: a highland zone (more than 2,000 meters) with high rainfall; an intermediate zone with some forest cover and steep slopes; and a lower zone of alluvial plains and seasonally flooded stretches of swamp and pasture (see Z. Abate, 1994). The basin now has a population of close to three million, but population densities here are much lower than in the highlands. Soils have not been as intensively exploited as in the highlands, and the abundant rainfall with low seasonal variation means that famine, unrelated to warfare, is rare. The steady seasonal flow of rivers means that hydropower generation can be maximized and the amount of water stored minimized.

Yet even in this most favored region, the real possibilities for "big project" development are limited and, as elsewhere, fraught with risks. It is a remote region. Even from the top of the escarpment, it is a long and challenging journey to Addis Ababa. From the floodplains it is even further. Indeed, up until the 1950s, the most efficient way to move goods out of the basin was by barge through the swamps, down the White Nile and to the railhead at Khartoum for ultimate shipment from Port Sudan.

In the so-called Gambella salient, jutting deep into the southern Sudan, the British, with the reluctant agreement of the emperor, set up a trading post and consulate (see Chapter 3). Here the British were able to dominate to a certain extent the trade in gold and ivory, and eventually coffee, that made the region's fortune. The emperors of Ethiopia tried, with some success, to redirect this trade through the highlands to Addis Ababa, Dire Dawa, and Djibouti.[11] However, the outbreak of civil war in the southern Sudan in 1955, and its nearly uninterrupted continuation ever since, meant that even the Sudanese trade exit was no longer available. The highland portions did move more and more of their coffee production eastward to Addis Ababa, while the lowland portions sank into torpor until the *dergue* decided that they could receive hundreds of thousands of refugees from the drought-stricken highlands after 1984.[12]

The region is one of highly mixed populations. In the upper altitudes the Oromo predominate. Some "Arab" Beni Shangul are present in the northwest sectors, while the lowlands are settled by Nuer, Anuak, and Shilluk. These black Nilotes are not always on the best of terms among themselves, and they have not tended to get along with the Oromo or the Beni Shangul. When high-land refugees were moved in large numbers into the lowlands, social and eco-logical equilibria were badly upset.[13]

The Akabo-Baro basin has virtually no industry, few urban centers, and scat-tered populations. Its demand for electrical power is tiny. Only "mini"-power projects of 1 to 10 MW make sense, so that while the region has enormous hy-dropower potential, it is too far from major centers of power consumption to warrant development at this time.

The upper altitudes, the zones of highest population density, have been the most intensively farmed. The steep pitch of slopes and fields, intensive farm-ing, and high rainfall mean that these areas are subject to heavy erosion. Large parts of the intermediate zones are also characterized by steeply pitched slopes (as in the northeastern rift valley, the "drop" from the top of the escarpment down to the floodplains is breathtaking). The lower alluvial plains are often flooded for five months of the year. The rivers emptying into the Sudd swamps of the southern Sudan often back up into the cultivable areas and pasture as the swamps seasonally expand. Cultivable and irrigable soils are frequently vertisols (as much as 60 percent in the lower areas), presenting all the problems already mentioned.

In most parts of the basin, given the long rainy season, what is really needed is supplemental irrigation when rains fail at the beginning or end of the grow-ing season. Yet it would not make economic sense to use surface irrigation sys-tems to deliver supplemental irrigation for low-value food crops. It might make sense for high-value export crops, but the difficulties of transportation and marketing have already been underscored. Tube wells might make sense in se-lect sites, but in the upper basin the water table is very deep below the surface.[14]

In the early 1980s, a Soviet team of consultants carried out a survey, includ-ing aerial mapping, of the basin, and identified several projects. Beginning in 1995, TAMS-ULG followed up, revising and expanding the database on river flows and site examinations. In its initial proposal of December 1995, the TAMS-ULG group identified seventeen projects. Several in the lower basin have very signifi-cant power ratings of 250 to 500 MW installed capacity. For the reasons stated above, the TAMS-ULG report did not find that development of this potential is warranted now.

In terms of irrigation, TAMS-ULG confirmed that there are some 1.1 million hectares of potentially irrigable land. About one-third is in the upper basin at altitudes above 425 meters, and two-thirds in the plains below 425 meters. The latter two-thirds are nearly all subject to seasonal flooding and backwater effects from swamp expansion, and they would require extensive diking to be suitable for irrigation. The only existing project is a small earthen dam on the Alwero, about 35 kilometers south of Gambella, intended to irrigate 10,000 hectares. It is safe to conclude that most development of irrigation will take place for the foreseeable future on the 330,000 hectares lying above 425 meters.

Thus, even this land of promise is a great distance in space and time from realizing its potential. Moreover, the possibility of social conflict as development occurs is very real. If one day the alluvial plains are irrigated and farmed, it is unlikely that the indigenous pastoralists and fishing communities will do the farming. More likely, as in the north, commercial farmers from the highlands may lease the irrigated land, assemble workforces from the highlands or from *fellata*, and marginalize the existing populations of Nuer, Anuak, and other so-called Nilotes. It has happened to the Afar in the Awash Valley, it is happening to the Hadendowa and Bani Amer in the northern Ethio-Sudanese border area, and it may well happen in the southwest salient.

CONCLUSION

One element of continuity linking Hailie Selassie, Mengistu Meriam, and the current head of state, Prime Minister Meles Zenawi, is a fascination, at least among top officialdom, with big projects. That fascination has been fueled by the international donor community, and, of course, by the contractors, equipment suppliers, and construction firms that live off it. The costs of such development are externalized to the local environment, local populations, and economic logic. Unlike China, Egypt, Brazil, or Turkey, Ethiopia has been too poor and too unstable to do much about its fascination. The Awash Valley Authority, with its explicit TVA contours, was the first (and so far the last) big Ethiopian project. Yet Ethiopia's very impoverishment, and its recent experience of severe famine, have made its policy-makers all the more determined to look for big solutions to big problems. They would do well to look closely at the Sudan's Gezira-Managil scheme, the world's largest agricultural scheme under single management as the Sudanese are wont to boast. It has not solved Sudan's needs in food production, nor in exports. It has drained investment away from other projects and other regions. Its own blueprint for organizing tenant farming collapsed years ago; the tenants, to use James Scott's image, reintroduced a production and social text that state planners could no longer read (1998:220). The infrastructure, without a cushion of export-generated foreign exchange, began to collapse in the mid-1980s if not earlier. In the past fifteen years regional famine has been no stranger to the Sudan.

A major problem is that while "we" (the donor community, environmentalists, academics, some policy-makers) know what went wrong, there is no compelling alternative blueprint that bureaucrats can embrace. The site-sensitive, hand-tailored, participative development recommendations elicit applause and sympathy but also bewilderment and hostility from public agencies that crave predictability, replicability, and easily monitored, standardized policies. Ethiopia will almost surely have a fling at a new generation of big projects. They may fail to achieve food security, but if pursued unilaterally, without the concurrence of Egypt and the Sudan, they will lessen the prospects for basin-wide cooperation.

Chapter 6 The Sudan: Master of the Middle

The Sudan enjoys two characteristics of overwhelming importance in the Nile basin. First, all major tributaries of the Main Nile—the White, the Blue, and the Atbara—flow through its territory and on to Egypt. The tributaries traversing the Sudan contribute to the flow of the main Nile in the proportions presented in Table 6.1. Second, it has by far the greatest potential for irrigated agriculture of any riparian, and its economic future cannot be conceived without exploiting that potential on a much greater scale.

The Sudanese economy continues to rely heavily on the agricultural sector. Moreover, it would take a near suspension of facts and belief to see industry, whether high- or low-tech, as the country's future. By contrast, its agricultural potential is tremendous. It has between 25 and 35 million acres of arable land, of which some 12 to 18 million are currently and, for the most part, primitively farmed. Only about 4 million acres are farmed under irrigation. There is abundant land of good quality lying between the White and Blue Niles. The Gezira-Managil scheme occupies a small wedge of this vast area. Because the Blue Nile basin lies at a higher elevation than does the White, gravity

Table 6.1 Contribution to Main Nile Discharge

	Annual	Flood Period
Blue Nile	59%	68%
Sobat	14%	5%
Atbara	13%	22%
Ethiopian sources:	86%	95%
Bahr al-Jabal (White Nile)	14%	5%
Total	100%	100%

Source: Adapted from Michael Field, "Developing the Nile." *World Crops,* 25, no. 1 (1973), pp. 11–15, and as in John Waterbury, *Hydropolitics of the Nile Valley* (Syracuse, N.Y.: Syracuse University Press, 1979), p. 23.

flow irrigation can be used to bring water to the lands between the two rivers. In addition, the soils are excellent, formed of alluvial deposits, and very deep. In sharp contrast to the Ethiopian highlands, they are relatively level so that irrigation perimeters require very little by way of preparation for irrigation and farming. The lowlands of Ethiopia, described in Chapter 5, form a continuum with the alluvial plains between the two Niles.

The Sudan could unquestionably double its irrigated acreage on a cost-effective basis, but might thereby add *25 million bcm or more* to its current water needs.[1] Even though Ethiopia sees itself in structural conflict with Egypt over Nile water, its long-term demands and potential drawdown on existing supply is not likely to exceed 4–5 bcm. The Sudan, which consistently portrays itself *officially* as in harmony with Egypt's Nile policies and acquired rights, is, in fact, in profound structural contradiction with its northern neighbor. There is an added element to the structural contradiction. Because the Blue Nile lies at a higher elevation than the White, it is in the Sudan's interest to pursue water storage and conservation projects on the Blue Nile. Using gravity flow for irrigation reduces the high costs of pumping water uphill from the lower-lying White Nile. By contrast Egypt has a long-standing interest in developing such projects in the Victoria, Albert, and Bahr al-Jabal sections of the White Nile, mainly because the riparians in these systems (Uganda and, de facto, the southern Sudan) are relatively weak.

During the negotiations for the 1959 agreement, and in various fora subsequently, Egypt has stressed that Sudan's agricultural potential in rainfed cultivation is great and should be exploited. It will be recalled that this is Egypt's message to Ethiopia as well. If one takes the gross volume of precipitation annually in the Sudan (or in Ethiopia), Egypt would seem to have a good case, but precipitation is seasonally concentrated, often too abundant to capture or store effectively, highly variable from one year to the next, and most abundant in areas, like the Sudd swamps, where cultivation is very difficult. No commercially viable agricultural sector anywhere has been able to prosper in the face of such patterns. Like Ethiopia, the Sudan has experienced drought-related famine, especially in the mid-1980s. War-induced famine in the southern Sudan is chronic.

If, as some predict, we see a secular increase in world prices for agricultural commodities, driven by the rising incomes of heretofore low-income giants like China, India, and Indonesia, the Sudan could clearly become a major net exporter of a range of such commodities, including coarse grains, medium- and long-staple cotton, oil seeds (sesame, sunflower), cane sugar, cattle, sheep, and hides. Outside agriculture and animal husbandry, the Sudan has potential in petroleum exports,[2] and it has long exported unskilled and semiskilled labor to Egypt and the Arabian peninsula. Oil exports could offer the Sudan another avenue to some prosperity, but as experience has shown elsewhere, oil rents often have a negative impact on the productive sectors of an economy.

If we could fast forward a generation or two, the prognosis might be considerably different, but we cannot skip the current generation in the Nile basin. In that light, it appears a virtual certainty that the Sudan's effective demand for Nile water will at least double.[3] If it does not, it will simply mean that the current economic collapse in the country continues.

UNSTABLE TRIADS

Bipolar conflicts, as in the Cold War, may produce a kind of equilibrium, but triadic conflicts are inherently unstable. The major conflict of interests in the Nile basin is, essentially, triadic, involving Egypt, the Sudan, and Ethiopia. Legro and Moravcsik (1998:60) put the "realist" case for this instability in stark terms: "If underlying state preferences are assumed to be zero-sum, there is generally no opportunity (*absent a third party at whose expense both benefit*) for mutually profitable compromise or contracting to a common institution in order to realize positive-sum gains. Nor can states engage in mutually beneficial simple political exchange through issue linkage. *The only way to redistribute resources is, therefore, to threaten punishment or to offer a side payment*" (emphasis added).

Unless one member of a triad is of overwhelming strength, that is, a true hegemon, then there will be a tendency for two to ally at the expense of a third party. Egypt is not militarily, economically, or geographically a pure hegemon. It has been able to make its interests prevail by suborning the Sudan in an alliance implicitly aimed at thwarting Ethiopia. The Sudan, out of objective economic interest and in hopes of extracting a better deal, might be tempted to ally with Ethiopia against Egypt. An Ethio-Egyptian alliance against the Sudan is likely only for short periods of time and for very specific purposes (for example, to contain or topple a specific Sudanese regime or government). Because any member of a triad may be tempted to shift its support within the group in order to extract a better deal from one of the two remaining protagonists, any equilibrium tends to be fragile and temporary.

A STABLE DYAD

It must be recognized, however, that since 1959 the Sudan has not broken ranks with Egypt, at least publicly, over Nile issues. Sudanese officials complain in private about the 1959 agreement and would like to revise it in favor of the Sudan, but vis-à-vis other riparians' claims to Nile water, the Sudan is as much a champion of the status quo as Egypt.[4]

This common cause was first put to the test in 1961, two years after the negotiation of the 1959 agreement. The East African riparians still under British control (Uganda, Kenya, and Tanzania), negotiated briefly with the Permanent Joint Technical Commission on the Nile (PJTC), the body designated in the 1959 agreement to represent Egypt and the Sudan vis-à-vis all third parties and to monitor implementation of the agreement itself. In the words of Paul Howell (1994:100–103), "It was admitted [by the PJTC] that they [the East African riparians] were in theory entitled to a share in the waters which originated in and passed through their territories, but only to water in excess of the requirements of the downstream states [Egypt and the Sudan], and there was none."[5] The PJTC put off East African demands by saying that anything could be discussed as needs arose. Ultimately Egypt rejected an East African claim for an allocation of 5 bcm as not supported by adequate data.

One of the more interesting areas of tacit cooperation between the two countries has come in Egypt's fairly routine release of water at Aswan in excess of its quota under the 1959 agreement. For example, it is estimated that between 1980 and the drought year of 1986 Egypt released in total 12 bcm in excess of its quota (Kliot, 1994:59). The excess water is the result of some combination of the Sudan's *under*consumption of its own share due to the prolonged economic crisis it has endured since the late 1970s, and of the fact that average Nile discharges have been above the average of 84 bcm used in the 1959 negotiations. The Sudan has watched silently as Egypt has routinely used around 60 bcm per annum in most years.

The stable front established by Egypt and the Sudan over the years has carried over into multilateral organizations and external funding agencies. Their position is not only taken as unassailable; it is sometimes argued more strenuously by third parties than by the two countries themselves. In 1989, an Economic Commission for Africa (ECA)/UNDP fact-finding mission, led by Roger Berthelot, is almost caricatural in its argumentation in support of the status quo in the basin. In an astonishing claim to exactitude, the report states in reference to Egypt and the Sudan (UNDP, 1989:3), "Under present circumstances, a water resources deficit in the amount of 4.8 bcm can be *precisely* forecasted by 2010" (emphasis added). It concedes that the riparians of the Lake Victoria Basin will need that much water themselves. Ethiopia was not included in the study. The report then enshrines the principle of "no net loss"—that is, for every cubic meter abstracted upstream, downstream flow should be enhanced by double that amount. Thus, if upstream abstraction rises to 5 bcm, then storage and channeling projects in the Sudd region must, in aggregate, reach 15 bcm, so as to deliver a net gain at Malakal (the northern outlet of the Sudd swamps) of 10 bcm.

The report sweepingly asserts (1989:5), "Under these circumstances, to satisfy the comprehensive water requirements of all the Nile countries, *there is no other solution* than appropriate regional water resources control based on upstream storage and joint water resources management" (emphasis added).

The report also assessed power needs throughout the basin and proposed an integrated grid, including the Inga power station in then-Zaire and far outside the basin, but not explicitly including Ethiopia. The inclusion of Inga legitimized the Egyptian project, worked out with its ally, President Mobutu of Zaire, to transport excess generating capacity at Inga eastward to the White Nile basin and then northward to Turkey and Europe. From the outset, Ethiopia viewed this project as expressly designed to preempt projects for the development of hydropower in Ethiopia capable of exporting power to other riparian states in the basin.

The report fails to consider any alternative scenarios. It does not entertain the possibility that Egypt and the Sudan may not be able to increase their utilizable supply or, further, that they may have to learn to do with less. Conversely, the report does not consider any upstream abstraction without compensation to the downstream riparians. Finally, it does not consider any reconfiguration of the irrigation infrastructure in Egypt and the Sudan to achieve greater efficiencies in water use. Straight-line projections from current use patterns prevail.

I have mentioned this report at some length because it came under the imprimatur of the ECA and the UNDP. It reflects the success of the solid Egypto-Sudanese dyad in propagating a vision of Nile cooperation that third-party funding agencies have, until recently, bought. The report, in this instance, was so one-sided that Ethiopia challenged its terms of reference and its conclusions. In light of Ethiopia's objections, the representatives of Kenya, Rwanda, Tanzania, Zaire, Uganda, and Burundi decided to refer Ethiopia's comments to their governments and to defer further discussion of the draft report.

The Permanent Joint Technical Commission on the Nile

The 1959 agreement provided for the establishment of the Permanent Joint Technical Commission on the Nile (PJTC). This is a remarkable creation, indeed an institution in the formal sense, that has continued to function almost without interruption through various political crises for nearly forty years. The PJTC has three principal functions. First, it provides for the reciprocal stationing in each country of engineers from the other to monitor the discharge at all storage sites to make sure they are in conformity with the basic allocation. I think it safe to say that reciprocity has been unequal, with Egypt's two dozen or so engineers in the Sudan taking a more hands-on monitoring role than their Sudanese counterparts at the Aswan High Dam.

As part of this function, the PJTC is to negotiate any reductions in the basic allocation brought about by prolonged regional drought. In the mid-1980s, that challenge was nearly posed, but both countries scraped by, adjusting summer and winter agriculture to reflect reduced supply. In the late 1980s, rainfall and water supply were restored to more normal levels so that the PJTC has not yet had to allocate a shortfall.

Second, the PJTC is to commission and supervise the engineering studies for and the actual implementation of any joint projects for water storage and supply enhancement. The primary focus here has been on the projects first identified under the Century Water Scheme and involving storage on the Victoria and Albert Niles coupled with the reduction of losses in the Sudd swamps. In this respect, after 1974, the PJTC commissioned the design and began the implementation of Jonglei I, a canal running on a south-north axis across the swamps to reduce spillage and channel water north to the White Nile at Malakal. The only project on Nile tributaries originating in Ethiopia was that at Khashm al-Girba on the Atbara. This project arose from the need to resettle Nubians whose villages were flooded by the reservoir upstream of the Aswan High Dam.

Third, the PJTC would represent Egypt and the Sudan in negotiations with any third party or parties, as it did in the 1961 negotiations with Britain's East African colonies (see above).

The permanent headquarters of the PJTC are located at Khartoum. The body meets quarterly, alternating between Khartoum and Cairo. There is no public chronicle of these meetings, but one may surmise that, for the most part, they involve the exchange of technical information.

Tremors in the Dyad

Egypt has had only three heads of state since 1952, and although Anwar Sadat was assassinated in 1981, the political establishment of Egypt has been notably unshaken. The same cannot be said of the Sudan. Its governments were changed extralegally in 1958, 1964, 1969, 1985, and 1989. In each instance relations with Sudan's large northern neighbor were affected. Nasser's death in 1970 also affected Egypto-Sudanese relations because the young Colonel Ja'afar al-Nimeiri, who had seized power in the Sudan in 1969, saw Nasser as his regional patron and his political model. Finally, with the exception of the period 1972–83, a civil war has raged in the southern Sudan. Through all this turmoil the PJTC has continued to meet regularly. Its one major joint effort, the construction of the Jonglei canal, fell victim to the civil war in the south, but not to differences between the two nations.

There was one interlude of high tension. In June 1995 there was an attempt on the life of President Hosni Mubarak in Addis Ababa, where he was attending a heads of state meeting of the Organization of African Unity (OAU). Egypt (and Ethiopia) accused the regime of Lieutenant Colonel Omar Bashir and Hassan Turabi of being behind the plot. Tensions between Egypt and the Sudan escalated. Egypt massed troops at a disputed border area at the Halaib triangle. The Sudanese government riposted by threatening to tear up the 1959 agreement and to alter the course of the Nile.[6]

Within months, however, relations between Egypt and the Sudan returned to something approaching normal. Indeed, in the spring of 1996, when the U.S. representative to the Security Council called for increased United Nations sanctions against the Sudan for exporting terrorism and abusing human rights, Egypt opposed them. The PJTC missed a few meetings, but by 1997 the situation had returned to normal.

The Egypto-Sudanese dyad is clearly not one between equals. Egypt effectively casts itself in the role of the Sudan's patron and protector vis-à-vis the outside world. Its toleration of the foibles and occasional hostility of different Sudanese regimes is simply testimony to its superior strength. For its part, the Sudan regards Egypt much as Mexico has traditionally regarded the United States. There has historically been a strong current of nationalist, anti-Egyptian sentiment dating back at least to the Mahdi and today manifested mainly in the Umma Party of Sadiq al-Mahdi. Yet *all* Sudanese leaders understand the superior power and reach of Egypt, and they feel uncomfortable and vulnerable if disputes between the two countries fester for very long. The Sudan has had to make the greater number of compromises over the years, a fact which provokes private resentment. Unequal compromise has been the price of the dyad's stability.

THE SUDAN'S NILE PRIORITIES

The 1959 agreement bore two immediate benefits for the Sudan and one mixed blessing. By accepting the construction of the Aswan High Dam, the Sudan in turn was authorized to construct the much-needed Roseires Dam upstream of the Sennar Dam, completed in 1926, on the Blue Nile. Roseires, completed in 1966, had a seasonal storage capacity of 2.4 bcm, and the water allowed the Gezira cotton scheme to nearly double its size. The newly irrigated area was known as the Managil extension, hence the Gezira-Managil scheme. A hydropower station was built at Roseires (210 MW installed capacity), and it became the primary source of electricity for the capital and its industrial zone. The second benefit was that the over-year storage capacity at the Aswan High Dam meant that Egypt was no longer concerned by "timely water," that is, the natural flow of the Nile in the months of low discharge prior to and including the spring and summer when Egypt's cotton crop was in the fields. After the construction of the Aswan High Dam, Egypt no longer cared *when* the water came from the Upper Basin as it could now be stored indefinitely and released when needed. This in turn meant that the Sudan was free to use stored water in the spring and summer.

The mixed blessing was the Khashm al-Girba Dam, completed in 1964 on the Atbara, and with a seasonal storage capacity of 1.3 bcm. This allowed the reclamation and irrigation of about 500,000 *feddan*s and the installation of a 15 MW power station. The downside was the forced resettlement of thousands of Nubians from northern Sudan, displaced by the Aswan High Dam, in a new and not altogether hospitable area.

Both dams immediately upon construction began to suffer severely from sedimentation. Khashm al-Girba lost half its storage capacity between 1964 and 1990 (Abdel Ati, 1992:32). Over the same period, the Roseires Dam lost a fifth of its capacity. Lost storage capacity reduced the amount of power that could be generated at both sites. Because of the sedimentation buildup, as little as possible of the silt-laden waters coming into the reservoirs in July and August are stored, thereby limiting summer cultivation and putting at risk cultivation during the peak months of October and November in years of low flood (Ahmed, 1996:95).

Like the series of dams in Ethiopia's Awash Valley, these new dams lead to a disruption of downstream traditional pastoralism and recessional agriculture without any compensatory mechanism for the affected populations (Abdel Ati, 1992:40; see also Abdel Nour and Mengestu, 1994:652).

By 1990 or thereabouts, the total capacity in *seasonal* storage of the three dams was at most 2.5 bcm. Contrast this with the over-year storage at the Aswan High Dam, which at the level of 175 meters asl is on the order of 118 bcm. This situation posed one immediate imperative for the Sudan, which, as with all else in the Sudan, proved not to be so immediate. This was to raise the height of Roseires from a designed retention level of 480 meters asl to 487 meters in a first stage, then to 490 meters in a second stage. This would increase the volume stored from the original 2.4 bcm to about 7 bcm. The need to undertake this project was identified by 1978 (Democratic Republic of the Sudan Ministry of Irrigation, 1978) and reconfirmed throughout the 1980s.

The Sudanese government has proceeded unilaterally in raising the height of the Roseires Dam without agreement with Egypt, let alone cost-sharing as provided for in the 1959 agreement. The Sudan is financing the project, estimated at about $400 million in 1989, out of its own meager resources. Work on Roseires began in 1994. Another seven projects, worth about $2 billion, have been studied and are ready for financing. All have a hydropower component, but official announcements have been circumspect in mentioning irrigation targets. Dams on the Seteit and the Atbara are planned, as well the Sabluqa Dam at the sixth cataract, Shareik at the fifth cataract, and Merowe at the fourth cataract on the main Nile (al-Moghraby, 1997; remarks by Minister of Irrigation Sayyid Ali Mohammed, *al-Hayat,* August 8, 1999).

Merowe has been long on the drawing boards. It lies at 298 meters asl. If ever built the dam would impound nearly 12 bcm, but it would also suffer from extremely high surface evaporation rates, leading at full capacity to the annual loss of some 1.7 bcm. The Merowe project would almost certainly be vetoed by Egypt, and, at an estimated $1.3 billion, is far beyond Sudan's financial reach.

There remains the vexed question of the Jebel Aulia barrage, built between 1932 and 1937, upstream of Khartoum on the White Nile. Its original purpose was to retain "timely" water for release to Egypt in the summer months, and its storage capacity is about 3.5 bcm. It is, in fact, a large artificial puddle with very high surface evaporation losses, as much as 2.5 bcm annually. It does not suffer from sedimentation, as the White Nile is relatively free of sediment (that is why it appears "white" in contrast to the sediment-laden "blue" Nile). Egypt no longer needs timely water, so, in theory, Jebel Aulia could be destroyed and some water saved. However, over the past sixty years a combination of private and cooperatively owned pump schemes have brought nearly 500,000 feddans under irrigated cultivation along the banks of the reservoir. No Sudanese regime has felt it worth the modest water gains to jeopardize these schemes. By contrast, some Sudanese water experts would like to divert water from the Jebel Aulia reservoir to irrigate land lying northwest of the barrage and which already contains abundant groundwater resources.

THE "BIG BLUE" TEMPTATION

If optimal water development were the overriding objective of Sudanese and Ethiopian policy-makers, then they would have achieved close technical cooperation and joint projects decades ago.

The U.S. Bureau of Reclamation study of 1964 may have been the first to suggest a large-scale joint project between the Sudan and Ethiopia. As with all parts of this study, the lightly hidden agenda was to send Egypt and the Soviet Union a message that the United States, through Ethiopia, could thoroughly disrupt Egypt's Soviet-funded water development plans. The proposal was for the so-called Border Dam, 20 kilometers upstream from the Sudanese border on the Blue Nile in Ethiopia. The bureau projected a reservoir of 11 bcm capacity in a fairly low-lying (ca. 500 meters asl), arid area, similar to the terrain around the Sennar and Roseires sites in the Sudan. Consequently, surface evaporation rates would be as high as at Roseires or Khashm al-Girba and so too would be sedimentation rates. The project was designed for power generation only (1,400 MW installed capacity), but there was no reason why it could not supply irrigation water by gravity to areas on both sides of the border (U.S. Department of Interior/Bureau of Reclamation, 1964; Guariso and Whittington 1987).

In 1984, the Bureau of Reclamation, in a study commissioned by the Sudanese government, invoked the Border Dam once again. While mentioning the challenge of arriving at a binational accord (at the time the Sudan was led by the pro-western Ja'afar al-Nimeiry and Ethiopia by the Marxist Menguistu Haile Meriam), the report argued that the Border Dam could solve the Sudan's growing power needs in the Khartoum, Wad Medani, and Atbara triangle, and provide additional irrigation water, thereby obviating the heightening of the Roseires Dam. The installed capacity at the Border Dam would be about five times the Sudan's current combined installed capacity at Roseires, Sennar, and Khashm al-Girba (U.S. Department of Interior/Bureau of Reclamation, 1984, V-10).

In confidential interviews with Sudanese and Ethiopian officials, I have been told that sporadic conversations were carried out over the years, but a combination of Egypt's likely veto and persistent political differences between Khartoum and Addis Ababa precluded any significant progress.

The Sudanese regime of Omar Bashir and Hassan Turabi determined that the overthrow of the *dergue* was in the interests of the Sudan and lent logistical support to the combined forces of the Tigrayan and Eritrean liberation fronts. Shortly after coming to power, representatives of the newly named Ethiopian People's Democratic Revolutionary Front (EPDRF) met with Sudanese officials in December 1992 and drew up an agreement on cooperation that included water issues. No mention was made of joint civil engineering works. Rather, stress was laid on watershed management, including reforestation of the Ethiopian watersheds, in order to reduce the sediment loads of the Blue Nile, the Atbara, and their tributaries.

H. E Hurst (1950:28) argued for another kind of border dam in Ethiopia. He noted that when the Baro River, which flows from southwestern Ethiopia into the Sobat (and is therefore part of the White Nile basin), is in flood over a five-month period, water spills over the floodplain with little water returning to the river when the water recedes. A solution, he suggested, would be to construct a dam of some 25 bcm capacity on the upper Baro in Ethiopia. He concedes that the appropriate site may not exist, and that even if it did, sedimentation might threaten it. Still, as we have seen in Chapter 5, Ethiopia, based on a master plan prepared by TAMS-ULG, may, through a series of dams on the Baro, provide the Sudan with just the sort of infrastructure to regulate flow that Hurst had advocated. This would, if realized, be a classic example of the unilateral provision of a public good. One party (Ethiopia) pays, but the second party (the Sudan) cannot be excluded from the benefits.

In sum, Ethio-Sudanese cooperation in the management of their shared basins is compelling. Egypt's priority for development of flow-enhancing projects in the White Nile basin is, at most, a second-best solution for the Sudan. Enhanced flow in the White Nile can be useful to the Sudan only through expensive pumping schemes that carry the water to higher elevations in the fertile plains lying between the two Niles.[7] White Nile projects are not well suited to power projects, given the gentle slope of the river, nor can they affect the sediment loads that have crippled the Sudan's seasonal storage sites on the Blue Nile and the Atbara. Despite the logic of Ethio-Sudanese cooperation, no coalition has ever emerged.[8] Were it to do so and remain stable, Egyptian economic and military power might be trumped.

THE FOURTH PLAYER: THE SOUTHERN SUDAN

The three-player game described above is substantially complicated by the fact that a strategic part of one player's territory, the southern Sudan, has not been under its effective control since 1955. It is much beyond the scope of this study to revisit the history of this civil war (but see Deng, 1995), and I will here touch upon only some salient features and events.

The main point is that so long as the ultimate disposition of this territory and its peoples remains undetermined, it is a bargaining chip for both Egypt and Ethiopia in their dealings with the Sudan. Neither country, however, knows how to use it. What is at stake is a third of the Sudan's territory, maybe a seventh of its population (no reliable census has been carried out in the southern Sudan, ever), and the Sudd swamps that control the discharge of the White Nile.

The struggle has been portrayed as racial and religious. Most of the peoples of the south are so-called Nilotes, negroid Africans who have populated parts of the Rift Valley and spread into east and central Africa. In the Sudan, they are composed of three large tribes inhabiting the swamp areas—the Dinka, the Nuer, and the Shilluk—and a myriad of smaller tribes in the equatorial highlands. The majority of the southerners are "animists," with significant Muslim and Christian pockets. The lingua franca of the south is a kind of pidgin Arabic. Population estimates are given by region in Table 6.2.

The first major break in the fighting came in February 1972 when Ja'afar al-

Table 6.2 The Regional Population
of the Southern Sudan, 1995

Bahr al-Ghazal	980,000
West Upper Nile	435,000
Lakes	665,000
West Equatoria	500,000
East Equatoria	730,000
Jonglei	530,000
East Upper Nile	120,000
Total	3,960,000

Source: UNICEF, *Operation Lifeline Sudan: Review of 1995 Activities* (Nairobi: UNICEF, 1995), p. 7, figure 3.

Nimeiri, only three years into his incumbency, accepted a peace agreement with the southern guerrilla leaders brokered by the World Council of Churches and by Emperor Hailie Selassie. The accords were negotiated in Addis Ababa and led to autonomy for the then three southern provinces of the Sudan. The federal government in Khartoum retained control of foreign policy, defense, and finances, while civilian southern ministries took control of education, health, agriculture, communications, police, and the like. Large numbers of guerrilla fighters were inducted into the regular Sudanese armed forces.

The agreement lasted for nearly eleven years. Once it was signed, the PJTC was able to carry out the studies and tender the projects for the Jonglei Canal, phase 1 or Jonglei I. The studies were begun in 1969, based on an earlier British survey over the period 1947–54 (the Jonglei Investigation Team). It was clear that the enormous rise in the levels of the equatorial lakes in 1961–64 and the consequent increased discharge downstream had fundamentally altered the regime of the swamps. It would be necessary to resurvey the entire area of the proposed canal through the swamps, but that could not be done so long as the civil war raged on.

With peace restored in 1972, the survey and design work began anew. In February 1974, Presidents Sadat and al-Nimeiri signed an accord that launched the Jonglei I project in earnest with costs and benefits to be shared equally. Construction began in 1978 and was suspended in 1983 when the civil war erupted once again in the south. About two-thirds of the canal's length had been excavated.[9]

During the sixty years of the Anglo-Egyptian control over the Sudan, the

British authorities governed the south apart. The goal was to limit the spread of Islam and of the Arabization of the populations. The tiny local intelligentsia tended to be educated in Christian parochial schools, and the few who could study abroad did so in the United Kingdom, in the United States, or at Makerere University in Uganda.

As independence for the Sudan approached in 1955, a mutiny broke out in the south among southern recruits protesting the introduction of northern officers. The fighting has gone on ever since with sporadic moments of truce and even one of reconciliation. The toll has been horrific. The south is dominated by the swamps that in turn make overland travel and communications enormously difficult. The few schools, clinics, and roads that the south has enjoyed have been repeatedly destroyed in rounds of fighting. Market agriculture has been nearly impossible, famine a frequent occurrence, and malnutrition and disease a fact of life and death. It would be hard to find any spot on earth where life has been so miserable for so long.

Yet the southern Sudan has always given rise to great expectations. None have been greater than those of David Hopper, writing when the first fruits of the Green Revolution were being gathered, when the term *sustainable agriculture* was virtually unknown, and when environmental issues had only begun to thrust themselves into the mindsets of the international donors (Hopper, 1976:201):

> Large areas of the Tropics are not farmed or grazed, and they constitute a huge reservoir of future production. The southern half of the Sudan is potentially one of the richest farming regions in the world, with the soil, sunlight, and water resources to produce enormous quantities of food—as much, perhaps, as the entire world now produces! *The water is useless today:* the headwaters of the White Nile, blocked in their northward flow by high plateaus, spill out over the land to form great swamps. To unlock the promise of the southern Sudan those swamps would have to be drained, a rural infrastructure put in place, and the nomadic cattle raisers of the region *somehow* changed into sedentary farmers. The capital costs of such an undertaking would be as large as the promise, and the time required would cover generations. Yet the potential is real and untapped, and as world food shortages persist such a reserve cannot long be neglected. (emphasis added)

One may pardon Hopper's extraordinary hubris, given his vantage point from the mid-1970s, but with such authoritative blessings from international experts, one may surely pardon the Egyptians and the Sudanese for dreaming similar dreams.

The Natural Regimes of the Sudd Region

By the time David Hopper wrote, strong critiques of the environmental impact of the Aswan High Dam had already appeared, and the first negative environmental analyses of the Jonglei I project were published (see Mann, 1977). Given current concerns with the protection of wetlands, it is easy to assume that any reduction in wetlands is by definition a bad thing. But with respect to the Sudd, we must understand the two very different natural regimes that have prevailed in the past half century: one before 1961 and another after 1964. During the years 1961–64 rainfall in equatorial East Africa increased dramatically, as did the discharge of the rivers that feed the Sudd swamps: the Albert Nile, the Bahr al-Jebel, the Bahr al-Zeraf, the Bahr al-Ghazal, and the Pibor-Sobat. The area of permanent swamps fed by these rivers may have increased roughly threefold from 7,000 square kilometers to 22,000 square kilometers. Temporary or seasonal swamps, fed mainly by torrents from the southwest watershed of the Sudd, increased on the same scale so that the entire swamp surface may have reached seasonally 40,000 square kilometers. This is about the size of Switzerland.

This expansion severely disrupted normal patterns of human and animal life, particularly in the central swamps through which the Jonglei Canal was proposed to be excavated (see Howell and Lock, 1994:253). Zeraf Island in the north-central swamps virtually disappeared. *Toich* (grass-covered mounds and tussocks that are accessible for grazing by cattle in the dry season when the temporary swamps dry up) lay permanently under water. Communications and overland transportation broke down. The only possible benefits of the new regime lay in the presumed increase of fish stocks and added protection against incursions from northern armed forces.

It may be the case that the region is gradually returning to the earlier regime anyway. There is inconclusive evidence of rising temperatures and declining rainfall in, for example, the eastern flank of the swamps (see Alvi, 1996). It is indisputable that Jonglei I would have a major impact on *any* natural regime. Howell and Lock (1994:257) estimate the impact of Jonglei I, assuming a throughput of, on average, 25 mcm per day, on three swamp regimes: 1950–61, 1961–80, and 1905–80. Taking only the latter, they estimate a loss of 27 percent of all *toich* land and 42 percent of the permanent swamp. Put in other terms, if we calculate that the Jonglei scheme will "save" 8 bcm from spilling into the swamps annually, this is enough water to cover 8,000 square kilometers to a depth of one meter.

The basic point is that Jonglei I could be implemented as planned without destroying the pre-1961 regime, and, as we shall see below, it would bring the local population several benefits. But Jonglei I would undo the regime established after 1964.

The Sudd Swamp Projects

Sir William Garstin first called for a canal to help channel floodwaters through the swamps to a northern outlet to the White Nile around Malakal. The pro-

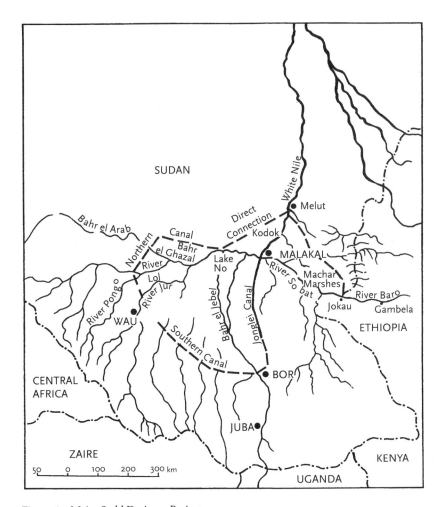

Figure 6.1. Major Sudd Drainage Projects
Note: The only partially completed project is the Jonglei Canal, two-thirds the distance from the Sobat River to Bor.
Source: Paul Howell, Michael Lock, and Stephen Cobb (eds.), *The Jonglei Canal: Impact and Opportunity* (New York: Cambridge University Press, 1988), p. 459.

posal was described, accompanied by marvelous aerial photographs, by Hurst and Phillips in the interwar years (see Hurst and Phillips, 1931) and studied in some depth by the Jonglei Investigation Team after 1947. The area was resurveyed in the mid-1970s and all potential water savings in the Sudd region were estimated (Table 6.3).

The Jonglei I project was begun in 1978, and consists of a 360-kilometer canal running from a point on the Bahr al-Ghazal about 40 kilometers to the west of where it joins the Sobat to form the White Nile, southward to Bor at the southern edge of the Sudd. By the time work was suspended in 1983, 267 kilometers of the canal had been excavated.

The presumed benefits of the canal were to be a reliable, year-round source of water, a reliable, year-round navigation channel for barges (thereby avoiding the shifting and vegetation-choked channels of the swamps), a year-round roadbed across the swamps, irrigation water for the surrounding areas,[10] a reliable water source for animal herds, and a source of modern sector jobs for the region's school population (for a recent statement of benefits, see Abd el-Ghany and Elwan, 1996).

The negative aspects of the canal, and they have been verified, are that it constitutes a major barrier to the movement of floodwaters from the rainfed eastern grasslands to the *toich* lands to the west of the canal (the so-called creeping flood). Although some transit points were provided for, the free movement of some 250,000 people and their 800,000 cattle along traditional transhumant routes would be severely impeded. Hutchinson (1996:352–53), in her study of the Nuer, shows that this is indeed the case.

Much earlier, in 1981, the Dutch firm ILACO sent agricultural survey teams that called into question the agricultural potential of the area. It is once again

Table 6.3 Gross and Net Water Savings from All Sudd Projects (in Billions of Cubic Meters)

	Increment at Malakal	Increment at Aswan
Bahr al-Ghazal	7.0	5.7
Jonglei I	4.75	3.85
Jonglei II	4.25	3.44
Sobat-Machar	4.0	3.3
Total	20.0	16.0

Source: Democratic Republic of the Sudan, Ministry of Irrigation, *Nile Waters Study,* vol. 1, *Main Report,* 1979, p.4.

(see Chapter 5) a question of vertisols that cannot be worked by machines when wet and which wreak havoc on metal blades and discs when dry.[11] Howell and Lock (1994:248) confirm this pessimistic assessment.

The other projects, first identified as part of the Century Water Scheme, have yet to begin, and I think it highly unlikely that they ever will. Jonglei II has generally been conceived as doubling the capacity of Jonglei I, from 25 mcm per day to 50 mcm. It has also been the general view that a second, parallel canal would have to be excavated. Further, in order to operate both canals, water would have to be stored somewhere further upstream. The Owen Falls Dam in Uganda does not have the capacity for such storage, and the favored site has always been Lake Albert, shared by the Congo and Uganda.

In March 1996 I interviewed a Sudanese water expert, Haydar Bakheit, seconded to the Tecconile headquarters at Entebbe, Uganda. He argued that the current, uncompleted Jonglei I should be expanded to have double its original capacity (with no parallel canal), and that storage at Lake Albert would not be necessary if the Bahr al-Jebel were embanked to prevent spillage and channeled directly into the expanded canal. The swamps, in his view, would be protected by the western torrents, rainfall (although my understanding is that surface evaporation in the Sudd region exceeds annual rainfall), and spillage in the Sobat and Pibor areas.

The other projects, long adumbrated, are the embankment and channeling of the Bahr al-Ghazal and of the Sobat, the latter greatly shrinking the Machar swamps of the east and the grasslands used by the local populations. The swamp regime would be deprived annually of some 20 bcm were all these projects to be implemented. Egypt and the Sudan would split some 16 bcm, as measured at Aswan, between them.

Even if these projects were carried out, their management would be complex and difficult to predict. Depending on the volume and velocity of each of the systems in spate (the Bahr al-Ghazal, the Jonglei Canal, or the Sobat), any one could force one or both of the others to back up, flooding their hinterlands. It is not clear if the White Nile, downstream of Malakal, could, because of its gentle slope, handle big seasonal increases in volume without flooding its hinterland.

Jonglei Politics

In 1974, when Egypt and the Sudan announced plans to go ahead with Jonglei I, there were protest riots in the southern capital, Juba. The Vice President for the Southern Region, Abel Alier, defied the rioters and said that Jonglei I would be vital for the progress and development of the south. The project went forward and undoubtedly generated local jobs and gave local workers, management, engineers, and others valuable skills.

The brief history of southern autonomy was stormy, and it must be said that Ja'afar al-Nimeiri did not play his hand in good faith. Three issues brought the experiment to a halt and led to a new and as-yet unfinished round of fighting. First, Khartoum subdivided the three southern regions into smaller units, thereby exacerbating intertribal rivalries. Second, oil was discovered at Bentiu, on the northern edge of the southern region. Khartoum refused any formula for sharing revenues, declared the oil fields outside the autonomous zone, and decided to transport the oil by pipeline to Port Sudan for refining or export as crude. Third, Ja'afar al-Nimeiri suddenly discovered his Islamic faith and established Islamic law (*shari'a*) throughout the land, including the largely non-Islamic south.

In combination, these moves rekindled southern resentment. Colonel John Garang deserted the Sudanese armed forces, returned to the south, and formed the Southern Peoples' Liberation Army and Liberation Movement (SPLA/SPLM). This is perceived to be a movement and guerrilla army dominated by the Dinka, which has allowed Khartoum to coopt southern leaders, like Riak Machar, a Nuer, to side with Khartoum against the SPLA. The war has sputtered on inconclusively since 1983, at great cost in lives and treasure to all sides.

One of the first acts in the new round of fighting was for the SPLA to close down the Jonglei I project and to destroy partially the sophisticated bucket dredger. It is easy to read from this act southerners' implacable hostility to the project in any form. I think that this is far from the truth. Garang took his Ph.D. at Iowa State University in 1981, and his thesis was on the socioeconomic development of the Jonglei area (Garang de Mabior, 1981). He was not opposed to the project per se, but rather suspected that the Egyptians and the northern Sudanese would reap all, or nearly all, of the benefits and reduce local, southern development to a cosmetic affair. In contrast, if Jonglei I were to be a growth pole for the south, and were the south to reap most of the benefits, then Garang was for it. This same view was echoed in an interview I conducted with a close adviser to Garang, Pagan Amum, in Nairobi in April 1996, and by several senior SPLM officials at a conference at New Cush in the southern Sudan a week later.

It is also the case that the SPLA/SPLM has sought not separation from the Sudan but rather the internal transformation of the Sudan into a secular, multisectarian, and multiracial entity. One may question the sincerity or the realism with which these views are espoused, but they are official. Thus, given that the

SPLM does not advocate the breakup of the Sudanese state, and that its leaders are not unambiguously opposed to Jonglei I, it is potentially an attractive partner or perhaps pawn for the Egyptians and the Ethiopians. If Egypt patronized the SPLM, it could become the arbiter of internal Sudanese politics, just as al-Nimeiri was the arbiter of southern politics from 1972 to 1983. However, Egypt doubts Garang's sincerity and fears being a catalyst to the breakup of an "Arab" state. Yet if there is no peace there will be no Jonglei I and hence no additional water for Egypt.

The Ethiopian government has no desire for an independent southern Sudan, which it sees as small, landlocked, desperately poor, and potentially burdensome for Ethiopia. It does not want to appear as the sponsor of the breakup of an African state, but having let Eritrea go, it has less of an ideological commitment than does Egypt. It is as concerned as Egypt about bringing the fighting to a halt, because the conflict could spill across the Gambella salient or anywhere along its long border with the Sudan, and because the civil war destabilizes the entire region. There is thus a three-party collective action problem, the solution to which is to find a formula to admit a fourth party to the table. There is one external body, IGADD, that has tried unsuccessfully to solve this collective action problem.

IGADD and the Greater Horn Initiative

In the wake of the severe droughts that gripped the Horn of Africa in the mid-1980s, a new organization, the Intergovernmental Authority on Drought and Development (IGADD), was founded in January 1986 and headquartered in Djibouti. It grouped Djibouti, Ethiopia (still under the *dergue*), Kenya, Somalia, the Sudan, and Uganda. Once it became independent, Eritrea was added. IGADD received funding and guidance from the FAO, the UNDP, the World Bank, the African Development Bank, and the International Fund for Agricultural Development. The initial purpose was to set up early warning systems to anticipate production collapses and potential famines, halt desertification, contain locust swarms, and promote food security. The member states contained 65 million farm families, and 11 million transhumant families, living in a zone with, on average, less than 400 millimeters of rain per year.

It was apparent that focusing on agricultural production systems, while important, could not on its own mitigate the effects of civil and tribal warfare that were afflicting the region. In the early 1990s, Somalia drew in UN peacekeeping forces, including U.S. troops, guerrilla forces overthrew the *dergue* in Ethiopia, and the civil war in the southern Sudan ground on. Because of the Somali de-

bacle, the United States became active on two fronts. The Administrator of
USAID, Brian Atwood, after visiting the Horn in early 1995, concluded that in
order to prevent future tragedies on the Somali model, preventive action was
required on both the political and economic fronts. Eventually Atwood was
joined by then-Vice President Al Gore in launching the Greater Horn Initiative. Part of its agenda was to mobilize the political leadership of IGADD to try
to broker a settlement in the southern Sudan. In this effort, Daniel Arap Moi of
Kenya was given the lead role. Simultaneously the OAU established a Conflict
Resolution Committee with similar objectives.

In 1995, IGADD had adopted a Declaration of Principles that included self-
determination for the southern Sudan. The Sudanese were rightly convinced
that IGADD was being manipulated by the United States, but because the Sudanese regime was so isolated internationally, it could not afford to reject this
forum in which it had full membership. IGADD-sponsored talks have limped
along without concrete results. U.S. interest in the Greater Horn Initiative
seemed to wane, especially as Ethiopia and Eritrea drifted into conflict, and
Uganda and Rwanda plunged into the Congo to try to oust Laurent Kabila. So
bizarre, even by the standards of East Africa, had been the reversals of fortune
and alliance,[12] that Eritrea's President Issayas Aferwerki signed a peace agreement with his erstwhile archenemy, President Bashir of the Sudan, in May 1999
in Doha, Qatar (*al-Hayat,* 3 May 1999, p. 5). Throughout the spring of 1999,
Egypt offered its good offices to bring about a settlement of the hostilities between Eritrea and Ethiopia. In mid-2000 Meles Zenawi of Ethiopia attended
the celebration of the "1989 Revolution" in the Sudan by an official visit to
Khartoum.

In sum, despite the somewhat unfocused and certainly unsustained efforts of
relatively rich and powerful third parties, no solution to the dilemma posed by
the war in the southern Sudan has been found. Ethiopia, Egypt, and the Sudan
are no closer than before to any mutual understanding. The two major public
goods at stake—peace in the Sudan and possible development of the southern
Sudd region—are, as usual, mirages on an ever-receding horizon.

CONCLUSION

The immediate structural conflict in the Nile basin involves a bipolar contest
between Egypt and Ethiopia. The long-term structural conflict involves a less
easily reconciled contest between Egypt and the Sudan. The Sudan's future is
hostage to developing irrigated agriculture on a scale that Ethiopia can never

contemplate. As has been the case since the mid-1970s, economic disarray and civil war prevent the Sudan from exploiting its agricultural potential. It is therefore always tempting for Egypt's adversaries to conclude that economic disarray and civil war in the Sudan are Egypt's doing. That suspicion is largely unwarranted, but I do believe that Egyptian policy-makers are divided as to whether it would serve Egypt's long-term interests to see the Sudan enter a period of stability and prosperity. Egyptian policy-makers must ask themselves the same question throughout the Nile basin.

The ultimate nightmare for Egypt would be for Ethiopia and the Sudan to overcome their domestic obstacles to development and to examine coolly their shared interests in joint development of their shared watershed in the Blue Nile, Atbara, and Sobat basins. Given Ethiopian and Sudanese regional behavior in the 1990s, Egypt need not lose sleep yet.

Chapter 7 Uganda: Egypt's Unwilling Ally

Uganda is a fourth, somewhat aloof, player in the main game among Egypt, Ethiopia, and the Sudan. It straddles the equator and enjoys abundant rainfall and a gentle climate. There are semi-arid regions in Uganda, especially in the northeast, and it could make far more extensive use of irrigation than it does currently (see below), but it is not overly concerned about water *supply*. It can prosper with what it has so long as it makes more efficient use of it. What shapes Uganda's interests in the Nile basin is the generation of hydroelectric power. At present, it has only one hydropower station, at Owen Falls, but there are a handful of sites on the Nile within its borders that could be developed in the future. To operate Owen Falls and any future sites requires maintaining relatively high surface levels of Lake Victoria and releasing large amounts of water through the turbines. Uganda's need for power meshes nicely with Egypt's need for water in the *White* Nile. The Sudan, as noted in the previous chapter, stands to benefit from development in the White Nile as well, especially through the first phase of the Jonglei Canal, but, if it had a free hand, the Sudan would prefer to invest in water control projects on the Blue Nile.

Thus, juxtaposed to the *potential,* or as yet unconsummated, natural alliance between the Sudan and Ethiopia, is the well-established wedding of Egypt and Uganda. Ethiopia and the Sudan have never explored fully their shared interests, partially because any such exploration would unquestionably antagonize Egypt. The rewards of this potential alliance would have to be very large indeed to offset the costs of Egyptian animosity. By contrast, Uganda has resented Egyptian hegemony, or at least high-handedness, in the Nile basin as much as any other riparian, but it has been locked into its alliance with Egypt by the two countries' objective, shared interests.

Schematically, we have the following alignment among the four key riparians:

For the Status Quo	For Change
Egypt	Ethiopia
Uganda	Sudan

THE LAKE VICTORIA REGIME

The constituent elements of Lake Victoria's water balance are not yet fully understood. Basic parameters, such as rainfall over the lake itself, remain the subject of debate. The single most contentious issue was and is the causes of the abrupt rise in lake level between 1961 and 1964. The Hydrometeorological Survey of the Catchments of Lake Victoria, Kyoga, and Albert, begun in 1967, was undertaken to answer this question but did not lay the controversy to rest (see Hydromet, 1974 and 1981).

Two explanations focus on the Owen Falls Dam, whose operation is jointly carried out by Egyptian monitors and Ugandan engineers of the Uganda Electricity Board (UEB). There was an illogical suspicion in the upper Nile states that Egypt provoked the rise in the lake level to store huge quantities of water upstream for subsequent release downstream in order to accelerate the filling of the reservoir at the recently completed Aswan High Dam. The problem with this theory is that, in the absence of the Jonglei Canal, additional water released upstream would simply spill into the Sudd swamps and for the most part be lost to evaporation. The second explanation, which has greater plausibility, is that the Ripon Falls, drowned by the reservoir upstream of the Owen Falls Dam, created a powerful backwater effect that prevented Lake Victoria from discharging excess rainwater in the early 1960s. In fact, in direct contradiction to the Egyptian conspiracy theory, part of the rock substratum that formed the

Ripon Falls had to be blasted away in 1959 in order to facilitate the discharge of Lake Victoria into the Victoria Nile (Wilson, 1967:6).

In the final analysis, it is abundantly clear that unusually heavy rainfall in the entire equatorial region is the main explanation for Victoria's sudden rise. All lakes in this part of the Rift Valley experienced similar rises (Kite, 1981), and subsequent attempts to simulate the lake rise through modeling concluded that increased rainfall provides all the answer one needs (Piper, Plinston, Sutcliffe, 1986). However, a crucial question remains unanswered: was the rise a cyclical phenomenon or merely an impressive blip in long-term trends? And if the answer is that it was a cyclical occurrence, how long is it likely to last? On the answer to these questions hinge the assumptions and parameters affecting all proposed projects to regulate the flow of the Victoria and Albert Niles and to generate additional hydropower.[1]

Another major parameter affecting Lake Victoria's water balance is runoff and inflow from the countries in its basin (Kenya, Tanzania, Burundi-Rwanda, and Uganda: I link Burundi and Rwanda because only Rwanda borders on the Lake but Burundi, through the Kagera River, contributes to the inflow into the lake). The Kagera River on average discharges between 6 and 7 bcm annually into Victoria while six smaller Kenyan rivers provide an equal amount.[2] Together these rivers provide over 70 percent of surface discharge into the lake, an amount equal to 50 percent of the annual throughput at the Owen Falls Dam. Any reduction in the discharge of these rivers could have serious consequences for any projects on the Victoria Nile or further downstream.

Surface evaporation at Lake Victoria averages about 1.4 meters across its surface area of 70,000 square kilometers. This is the equivalent of nearly 100 bcm or considerably more than the average flow of the Nile at 84 bcm. Similar or higher rates of surface evaporation are found in the other equatorial lakes, and the problem is exacerbated by the rapid expansion of water hyacinth, which can double or even triple the rate of surface evaporation (Ntale, 1996). Even if the rainfall regime were to remain steady for some time, reductions in inflow through more intensive agricultural use and storage for power generation, coupled with accelerated surface evaporation through the spread of the water hyacinth, could lead to a significant drop in lake levels and to reduced discharge at Owen Falls Dam.

COOPERATION AMONG
LAKE VICTORIA RIPARIANS

In the space of thirty years, Great Britain, the former Nile basin hegemon, managed to take directly contradictory positions on the proper regime for the use of the river. In the 1929 agreement between Egypt and the Sudan, both under British suzerainty at the time, protection of Egypt's acquired rights, and secondarily those, newly established, of the Sudan, was strongly emphasized (as cited in Water Development Department, 1993:3): "Save with the previous agreement of the Egyptian Government, no irrigation or power works or measures are to be constructed or taken on the River Nile and its branches or on the lakes from which it flows, so far as all these are in the Sudan or in countries under British Administration which would, in such manner as to entail any prejudice to the interests of Egypt, either reduce the quantity of water arriving in Egypt or modify the date of its arrival or lower its level." Three decades later, after Egypt and the Sudan had become sovereign states and at the time they were negotiating the 1959 Agreement Between the Republic of the Sudan and the United Arab Republic for the Full Utilization of the Nile, Great Britain reversed its stance and implicitly advocated the position we now associate with the "community of users" and with equitable use:

> The territories of British East Africa will need for their development more water than they at present use and will wish their claims for more water to be recognized by the other states concerned. Moreover, they will find it difficult to press ahead with their own development until they know what new works the downstream States will require on the headwaters within British African territory. For these reasons the United Kingdom Government would welcome an early settlement of the whole Nile Waters question. A conference of all riparian states has been suggested. In principle the United Kingdom Government favor this idea but think that a conference is unlikely to be successful until the Sudan and the United Arab Republic have settled the difference between them. (Overseas Press Service, August 27, 1959, as cited in Waterbury, 1982:125)

At the time, Great Britain thought that its East African territories would need about 2 bcm to meet their irrigation and power needs. The East African states themselves rejected this figure and in 1961 presented to the Permanent Joint Technical Commission on the Nile a request for 5 bcm. As noted in Chapter 6, this request was rejected as not being supported by adequate data (Howell, 1994:103).[3]

I have stressed the unremitting political and economic turmoil that characterized the upper basin riparians from the early 1960s virtually to the present time, with some relative stabilization occurring after 1990. Projects that might have made claim to some of the 5 bcm requested in 1961 languished for want of political attention and financing. Even today, any estimate of the real demand for irrigation water derived from sources of the Nile is guesswork. With some caution, therefore, one may advance the figures in Table 7.1.

I believe that at a *maximum* these countries might reasonably expect to irrigate 1 million new hectares of land, with another half million hectares of supplemental irrigation. With a very large margin of error, one may estimate average water use per hectare at 10,000 m^3 per year, for a total of 10 bcm (see Kagera Basin Organization, 1979, vol. 13:35; UNDP, 1989: chapter 4, pp. 15–17; Mbugua, 1987:45; Kabanda and Kahangire, 1991:4).

This figure could be much lower if Tanzania fails to implement the so-called Smith Sound project. This goes back to the period of German control of Tanganyika and was further studied by the British. At present it would consist in excavating a 600-kilometer canal from Smith Sound on the south shore of Lake Victoria to the central plains of Tanzania in the Dodoma region, where some 600,000 hectares might be brought under irrigated cultivation. Tanzania still has this project under consideration, but the bleak prospects for financing it, at an estimated $6 billion, preclude any progress in the foreseeable future.

Uganda would face a more severe threat than any other Nile riparian if Kenya, Tanzania, Burundi, and Rwanda were to abstract significant amounts of Nile water for irrigation because that conceivably could drop the level of Lake Victoria and reduce power generation at Owen Falls, thereby rendering uneconomical other Ugandan projects to increase power generation. Egypt and the Sudan would not be pleased, but because so much water is currently lost in the Sudd swamps, a reduced flow from Lake Victoria through Owen Falls Dam would probably have little effect on the *current* (i.e., in the absence of Jonglei I or II) discharge from the swamps at Malakal.

Subsequent efforts to engineer a coordinated stance among the Lake Victoria riparians were initiated by Egypt and consisted in the launching of Hydromet in 1967, then the Undugu Group in 1983, and finally the Tecconile project

Table 7.1 Potential Irrigated Hectareage Drawing
on Nile Sources in the Lake Victorian Basin

Country	Hectareage
Kagera Basin	200,000 new
	300,000 supplementary
Kenya	200,000
Tanzania	600,000?
Uganda	200,000

in late 1992.[4] Because Egypt was seen as the originator of these organizations, the upper basin riparians signed on symbolically or took observer's status (Kenya and Ethiopia). Beginning in 1997 (see Chapter 3) the World Bank and the UNDP essentially assumed responsibility for brokering a new regime through the Nile basin Initiative.

There have been only two initiatives originating among the Lake Victoria riparians themselves. The first was the establishment of the Kagera Basin Organization in 1977, and the second was the Lake Victoria Environmental Management Project of 1996.

The Kagera Basin Organization

The Kagera Basin Organization (KBO) is, on paper, one of the most ambitious and coherent river organizations in Africa if not the world. Four states share the Kagera basin: Tanzania (35 percent of the watershed), Rwanda (33 percent), Burundi (22 percent), and Uganda (10 percent). The river, as noted, annually discharges between 6 and 7 bcm into Lake Victoria. It is fed by the Ruvubu River arising in Burundi, the Kanyuru, a border river between Rwanda and Burundi, and the Nyabarongo in Rwanda. The basin enjoys relatively high and even rainfall. Irrigation per se is not a major priority for the KBO, although power generation is.

In 1967–68, with some prodding by Belgium (formerly responsible for Rwanda and Burundi), the EU, and the UNDP, the four riparians reached a technical agreement for the study of the development of the Kagera basin. Phase I of the comprehensive survey of the basin began in 1971, at the same time that Idi Amin pulled Uganda out of the emerging group. A decade later, after Idi Amin had been driven from power with the help of the Tanzanian military, Milton Obote brought Uganda back into the KBO.

The Kagera River Basin Development Plan was completed in 1973 and resulted in the identification of a number of sectoral follow-up studies. On August 24, 1977, Rwanda, Burundi, and Tanzania signed the Rusumo Agreement, giving birth to the Organization for the Management and Development of the Kagera River Basin, which was formally launched on February 5, 1978. The leitmotif of the KBO was "integrated basin development." Sectoral studies dealt with water, agriculture, mineral exploitation, disease and pest control, transport and communications, trade, tourism, wildlife, fisheries, industry, and environmental protection. This is a blueprint for a multi-good cooperative framework, where bargaining, compromise, and compensation could be worked out over the range of sectors and objectives contained in the integrative development approach (see KBO Secretariat, 1979; World Bank, 1979; Okidi Odidi, 1986). It is precisely this sort of multi-good context that Egypt habitually sought to promote in its dealings with the upper basin riparians, particularly in the context of the Undugu Group (see Chapter 3).

The keystone of the KBO was to be the dam and power station at Rusumu Falls. It would involve a reservoir some 200 kilometers long with a capacity of about 13 bcm. Tanzania initially wanted a power station with 130 MW capacity, but because of the extent of potential flooding such a power station and its reservoir would require, it was scaled back, in 1981, to 80 MW.

The institutional framework for the KBO was designed by two eminent international lawyers and experts in transboundary resources issues: Guillermo Cano of Argentina and Roger Hayton of the United States. The two pillars of the KBO were the Commission, an executive organ comprised of representatives of each member country that met three times annually, and the Secretariat. The KBO was endowed with a legal personality allowing it to enter into contracts, acquire and dispose of movable and immovable property, and to be a party to legal proceedings. It could enter into agreements for technical assistance and for project financing. Between 1978 and 1981, the KBO budget was supported by Burundi (25 percent), Rwanda (35 percent), and Tanzania (40 percent). When Uganda rejoined in 1981, each riparian was obligated to pay one-quarter of the budget. In short, the institutional arrangements provided for considerable autonomy for the KBO Commission and its Secretariat. It was constituted to act as a true development authority for the basin in the same sense as the Tennessee Valley Authority.

Nonetheless, the arrangements on paper could not solve the collective action problem. Domestic politics ultimately were trump. Uganda descended once again into civil war after 1986, as Yoweri Museveni slowly fought his way to power at the expense of Milton Obote. Tanzania drifted into economic crisis and the final discrediting of Julius Nyerere's African socialism. A few years later Rwanda dissolved into the attempted ethnocide of Hutus against Tutsis, which brought the Kagera River briefly to world attention as thousands of bodies washed down its course into Lake Victoria. The KBO still exists, and its headquarters remains at Kigali in Rwanda. The Rusumu Falls project exists only on paper.

The Lake Victoria Environmental Management Project

Lake Victoria is the second largest body of freshwater in the world (69,500 square kilometers) after Lake Superior. Yet like the U.S. Great Lakes, it is increasingly subject to deteriorating water quality and eutrophication. Coastal towns and cities pour untreated sewage directly into the lake, while fertilizer and pesticide runoff from coastal farms and from rivers is on the increase. By far the most serious threat to Victoria and the other equatorial lakes is the rapid spread of the water hyacinth.

The water hyacinth is an ornamental plant that originated in Latin America. It may have made its way to Africa and elsewhere (it is present in more than fifty countries) clinging to ship bottoms, but it has also been released into waterways by individuals who have used it to decorate garden ponds and basins. It has been present in Egyptian irrigation canals and the Nile itself since the turn of the twentieth century. Further upstream, and as early as 1960, large islands of hyacinth were lodged against the Jebel Aulia Dam in the Sudan, and by 1975 it was estimated that 3,000 square kilometers of the surface of the Sudd swamps were covered by the plant (Waterbury, 1979:237). Even though the plant may have been transiting the Upper Nile system for years, it did not take root, as it were, until the mid-1980s in Lake Kyoga (Uganda) and in Lake Victoria itself in 1990 (FAO, 1995).

Water hyacinth is a scourge with no redeeming features. It thrives on the increasing effluent discharge into equatorial lakes and watercourses. It chokes harbors, impedes fishing boats, reduces oxygen and hence fish stocks, doubles or triples surface evaporation rates, sinks and rots as entire islands of vegetal mass submerge, and now threatens the turbines at the Owen Falls Dam.

The hyacinth defies mechanical harvesting, and even when harvested it has no practical use. Animals will not eat it, it cannot be composted into anything useful, and it cannot be transformed into combustible fuel at any acceptable cost. Uganda suffers disproportionately from this scourge. Prevailing winds over Lake Victoria blow islands away from Kenyan and Tanzanian shores and toward Ugandan shores and ports (Port Bell, Entebbe, Jinja). At Owen Falls the pressure exerted by the Victoria Nile on the hyacinth mass, estimated at 70,000 tons and pressed against the booms, occasionally sinks large islands that then jam one or more turbines forcing their shutdown (Agricultural Policy Committee, 1995). The mass, kilometers in length, is being attacked manually by demobilized Ugandan soldiers who are exposed to snakes and rodents that thrive in and on the hyacinth islands.[5] The hyacinth contributes to what at least one author sees as the rapid death of Lake Kyoga (Ntale, 1996).

Owing to the hyacinth, coupled with discharge of effluents and agricultural run-off, the World Bank, in 1996, allocated $50 million to combat the plant and to secure the establishment of the Lake Victoria Environmental Management Project in which all lake riparians have agreed to participate. Because Uganda suffers from the hyacinth more than the others we may expect some free-riding by Kenya, Tanzania, and Rwanda.

The means of combat at the disposal of these states are untested, and all may have serious side effects. Florida has been successful in using a herbicide that degrades quickly in water and that does not harm fish life. The risk here is that killing off large masses of the hyacinth may cause them to sink and decompose, thereby exacerbating sedimentation and eutrophication. No mechanical harvesters have been up to the task of containing the hyacinth masses in the lake system. Natural predators or feeders, such as an Amazonian weevil, can be effective only after substantial chemical and mechanical control has reduced the plant mass on a large scale. Nonetheless, all three methods will have to be brought into play within the inter-riparian framework established in 1996 (see FAO, 1995; Kaufman, 1995; Ministry of Natural Resources, 1995).

UGANDA'S OPTIONS

Virtually all of Uganda lies within the Nile basin or drainage area. No other riparian can make that claim. Virtually every drop of water it uses is technically Nile water. If dependency on the resource is a heavily weighted factor in determining equitable use, then Uganda has equity on its side. At the same time, with the exception of the semi-arid northeast of the country (Karamoja), Uganda has abundant rainfall and very little surface irrigation—only 30,000 hectares in the late 1990s. Although the fao in 1987 put forth a heady estimate of potential irrigated hectareage at 410,000, a more reasonable figure is 190,000 (UNDP, 1989:13; Kahangire and Dribidu, 1995). Uganda's Ministry of Natural Resources, in its Water Action Plan (1995:3), was cautious: "It is difficult to predict the trends of agriculture in the country—and therefore to be able to estimate the demands irrigation will put on surface water sources. The economic and social feasibility of large scale irrigation compared with alternative agricultural practices is not obvious—so demands for irrigation water are not likely to be significant in the near future." The same document endorsed a brave principle that may be exceedingly difficult to apply: "Water should be considered as a social and economic good, with a value reflecting its most valuable potential use."

Uganda's most pressing need is for hydropower, and its potential is great. The drop of the Victoria Nile from its outlet at Lake Victoria to Lake Kyoga is 105 meters, and from Lake Kyoga to Lake Albert is another 410 meters. There is the potential to generate at least 2,000 MW along both stretches, yet actual installed capacity is less than 400 MW. Thus, Uganda has no urgent plans to abstract large amounts of irrigation water, whereas pursuit of hydroelectric schemes will place a premium on maximizing the flow of the Victoria and Albert Niles. That, as already stated, makes Uganda a natural if unwilling partner of Egypt, and, if anything, a more concerned riparian with respect to any abstractions the other Lake Victoria states may be contemplating.

This has led to a kind of schizophrenia among Ugandan policy-makers and intellectuals. John Ntambirewki, a professor of international law, and Akiki Mujaju, a political scientist, wrote a commissioned paper for the Water Development Department of the Ministry of Natural Resources (1993). In it they categorically rejected any continued binding effect from the 1929 agreement. The principles enforcing Egypt's acquired rights in that agreement were explicitly endorsed in the 1949 agreement between Great Britain, acting on behalf of Uganda, and Egypt for the construction of the Owen Falls Dam. In effect, the two authors portrayed the 1949 agreement as having been imposed on Uganda. They point out that the 1949 agreement called for review and revision twenty years after storage at Owen Falls had begun. When that period came up, Idi Amin was in power and not paying attention, so there was no review or renegotiation. On that and other counts, the authors argue, the 1949 agreement is no longer binding. They then lay out the prevailing image of Egyptian intentions: "What stands out clearly from the foregoing exposition is the attempt by Egypt to dominate the Nile Valley, first as an imperial power which failed, and secondly as a surrogate of an imperial power. In its latter position, exploiting the might and sympathy of Britain, Egypt sought by agreements to achieve what it had failed to do by naked conquest, the complete control of the Nile waters for its exclusive use" (1993:6).

Both John Ntambirewki and Patrick Kahangire, head of the Water Development Department, privately recognize that renegotiation of the 1949 agreement is meaningless unless Uganda truly feels that the regime established at Owen Falls through that agreement is not in its interest. Neither believes that that is the case. As Kahangire stated to me, "Why renegotiate or review Owen Falls if we don't know what we want changed? We have an interest in the curve [see below] as stated" (November 17, 1995). Ntambirewki singled out Kenya as the single largest obstacle to cooperation among the upper basin riparians. He went on: "In a fundamental way our strategic alignment should be with Egypt and the Sudan. Our overwhelming objective is maximum flow through Owen Falls. Kenya is our real problem" (November 14, 1995).

Ugandan Power Schemes

Only two projects included in the Century Water Scheme have actually been implemented, and one, the Jebel Aulia barrage in the Sudan, is now obsolete (see Chapter 6). The other, the Owen Falls Dam, has, since its completion in 1954, served a dual purpose: it regulates the outflow of the Victoria Nile from Lake Victoria in the interests of Egypt, and it generates power for Uganda. The power supplies the capital at Kampala and the industries around Jinja, and part is exported to western Kenya with smaller quantities to the Bukoba region of Tanzania.

The 1949 agreement between Egypt and the United Kingdom acting on behalf of Uganda stipulated that operation of the Owen Falls Dam would be according to an "agreed curve" that would, based on ten-day average flows, simulate the natural flow of the Victoria Nile. Egyptian engineers were to be stationed at Jinja, with direct access to a gauge two kilometers upstream of the dam, to verify that the curve was being respected. The engineers are there today, and the gauge, in a raised platform shed, is as well. The Ugandan officials of the Uganda Electricity Board, which operates the dam, provide the Egyptians with detailed daily flow data through the turbines and through the sluices.

When the agreement was concluded, the Aswan High Dam project was not even under consideration. Egypt, in 1949, was concerned to use Lake Victoria as a seasonal storage site so that "timely" water could be released during the period when the main Nile normally runs low. For that reason, Egypt, through a codicil to the 1949 agreement, obtained Uganda's consent to raise the dam one meter higher than necessary for power generation. Egypt assumed the entire construction costs of the additional one meter and paid Uganda a one-time compensation for forgone power of £980,000 (Wilson, 1967:5). The agreement granted Uganda, so Ntambirewki and Mujaju claim (1993), a contingent sovereign right to use its water as it sees fit so long as it does not entail any prejudice to Egypt's interests in accordance with the Nile Waters Agreement of 1929.

The original power station was commissioned over the period 1954–68 with the installation of ten power sets of 15 MW each for total installed capacity of 150 MW. During that period, the level of Lake Victoria rose dramatically, and, according to the agreed curve, water release at Owen Falls rose sharply as well. It was also during those years, with the construction of the Aswan High Dam, that Egypt lost interest in using Lake Victoria to store "timely" water as the Aswan High Dam took over that function. Uganda thus could try to maximize power output at Owen Falls so long as the agreed curve was respected. However, this was only a technical possibility as Uganda entered a long period of political disintegration and economic decline precipitated by Idi Amin and continued after 1981 by the return of Milton Obote. It was not until the late 1980s, under Yoweri Museveni, that Uganda began to get a grip on its economic fate.

In 1989 the Uganda Electric Board contracted with Acres International of Canada to carry out a feasibility study of an extension of the Owen Falls power station (see Acres, 1990). The extension would add 102 MW capacity to the existing 150 MW. It would allow Uganda to make use of the additional, average throughput of 400 cubic meters per second that had characterized discharges at the dam since 1964.[6] The extension, costed at about $300 million, would be economical only if the high rates of discharge were maintained throughout the long, projected life of the project. The Acres report and subsequent World Bank reports argued that the averages established since the early 1960s were likely to last.

The Acres report questioned whether the pre-1961 flows had been accurately measured and suggested they had been underestimated. The World Bank, in recommending the project for funding (see World Bank, 1991:65) adopted that view and put the probability of reverting to the low flows, pre-1954, at 1 percent and the probability of continuing the high flows, post-1954, at 99 percent. Carbon dating at Hippo Bay near Entebbe suggest that the lake levels reached after 1961 had not been achieved in the preceding 3,700 years (see note 1), prompting one skeptic to query "should we not consider as 'normal' what has prevailed for 3700 years as opposed to what has prevailed for 30?"

The Acres Report became justifiably controversial, even to the extent that the World Bank quietly sought a reevaluation of its findings. Other agencies questioned the conclusions as well. Britain's Overseas Development Administration (ODA) issued a report in 1993 (Overseas Development Administration, 1993). This report does not shout the emperor is naked, but it comes close. It refers to studies of the Institute of Hydrology from 1983 to 1985, and a subsequent ODA-commissioned study through Sir Alexander Gibb, that came to significantly different conclusions than Acres. It is noted that over the twenty years prior to 1990 more intensive agricultural and municipal use of water may have reduced gross inflows into Victoria by 10 percent and that that rate is likely to continue at about 1 percent per annum (1993:44). Further, the average difference between rainfall and surface evaporation is quite narrow, so that small changes in rainfall can cause big changes in outflow—in either direction. Third, given the vast volume of water stored in Victoria, it may take several years for these net changes, in either direction, to manifest themselves in out-flow. The Institute of Hydrology study estimates about a 25 percent chance of the pre-1961 rainfall and outflow regime returning (1993:66). It is the fact that a sharp decline in rainfall in 1993–94 lowered the lake level by about one meter, although in the years immediately following it partially recovered. There are those, nonetheless, who believe that there is a secular downward trend in rain-fall toward means established before 1961.

If there is this much room for doubt, why is the extension going forward? A possible explanation is that by 1988–89 Uganda had already established rapid rates of GDP growth. The major bottleneck to continued growth was seen as in-adequate power supplies both for domestic use and for export. The Uganda Electricity Board, which is the commissioning agency for hydropower projects, saw the extension as the lowest-cost quick fix to the problem. The UEB does not have sophisticated in-house hydrological expertise, and it has allegedly been re-luctant to reach out to those agencies that have such capacity. At the same time, the World Bank was eager to do something for a new African regime that was boldly pursuing structural reforms and achieving high rates of growth. None of the principals in this project wanted to hear bad news, and Acres conveniently gave them none.

Not only might the economic feasibility of the Owen Falls extension be jeopardized if the skeptics are right; so too would the Bujagali project, launched in February 1996. Bujagali is a spectacular set of rapids about 7 kilometers downstream from Owen Falls. The project is a private sector venture, allowed through changes in Ugandan law brought about in 1995. The main actor is Madhvani International, S.A. This international holding company is owned by members of the Madhvani family, prominent Asian businesspeople who had been expelled from Uganda by Idi Amin and welcomed back by Yoweri Musev-eni. The operating company is Nile Independent Power (NIP), and the consult-ing engineers are AES of Arlington, Virginia. The agreement to launch the pro-ject was signed on February 23, 1996, by Secretary of Commerce Ron Brown, shortly before his death, with representatives of the UEB and NIP. The power plant will have an installed capacity of 340 MW. It will sell bulk power to the UEB, which will be responsible for distribution. Some 200 MW will be ex-ported to the other Lake Victoria states. This is the first private power project on this scale in sub-Saharan Africa and is testimony to the possibilities for pri-vately financed power schemes in countries with severely constrained public in-vestment capabilities.

If, however, the discharge regime at the outlet of the Victoria Nile were to revert to that of thirty-five years ago, then both the Owen Falls extension and the Bujagali scheme would not be able to cover their costs.

Lake Albert and Jonglei

To date Egypt and the Sudan have sought from Uganda only that it not interfere appreciably in the natural regime of the Victoria and Albert Niles. However, the day may come when Egypt, in particular, seeks a formal understanding with Uganda to tamper quite spectacularly with the natural regime of the Albert Nile (see Hurst, Black, and Simaika, 1966:15–16). In the previous chapter, I have already explored the basic outlines of the Jonglei I and II canal projects to reduce spillage and surface evaporation in the Sudd swamps of the southern Sudan. Only were Jonglei II to become a real possibility would Uganda be approached for a cooperative understanding. So too would the Congo.

Lake Albert (for a time known also as Lake Mobutu Sese Seko) is shared by Uganda and the Congo. Neither country uses it for water supply, but it is an important source of fish. Parts of the Albert Nile, which flows north to the Sudanese border, are navigable, although water hyacinth is increasingly interfering with both fishing and navigation.

As the pieces of the Century Water Scheme were being put together, Lake Albert was seen as an ideal storage site for water to be released during the "timely season" of low water in the main Nile. Releases of water during the dry season in the swamps would reverse to some extent the natural regime of the river. By channeling these releases through the proposed Jonglei I and II canals, spillage in the swamps would be significantly reduced. What makes Albert attractive is its surface-to-volume ratio—that is, relatively small surface to relatively large volume. This is a function of its relatively steep sides. After World War I, a study carried out by Sir Murdoch MacDonald (Ministry of Public Works, Egypt, 1920:132–33), noted that a dam about 50 kilometers downstream of the outlet of the Albert Nile could raise the lake level by 7 or 8 meters, thereby increasing the volume stored by 40 bcm. This could be achieved, so it was believed, with minimal flooding of inhabited areas and minimal disruption to human activities.

The sharp rise in lake levels brought about by the heavy rains of the early 1960s was an unplanned test of this hypothesis. Prior to 1960, the average Lake Albert level was 620 meters asl or 10.6 meters on the principal gauge at Butiaba. At that level, lake surface was 5,700 square kilometers and lake volume 155 bcm. By June 1963, the level of Albert had risen nearly 4 meters, increasing the surface to 6,100 square kilometers and the volume to 180 bcm. In order that Jonglei I and II function optimally, the level might have to go as high as 640 meters asl (30 meters on the Butiaba gauge) with storage of 280 bcm (Hydromet, 1974: vol. 1, pt. 2:903; Haynes and Whittington, 1981).

The drawbacks for Uganda are that at that level something on the order of 1,000 square kilometers of land would be flooded, particularly along the eastern shore of the lake. The backwater effect on the stretch of the Nile running from Lake Kyoga to Lake Albert would be significant and might impact negatively on Murchison Falls. In addition, the slope of the Albert Nile is quite gentle so that increased releases from Lake Albert might contribute to flooding. The principal gain for Uganda would be power generation at the site of the regulator dam. Egypt might also have to consider paying Uganda annual compensation or rent for the facility. There is no gain at all for the Congo, which would surely seek financial compensation of some sort as the price for its cooperation. The Sudan, I think it can be safely predicted, will see this project as Egypt's affair. Jonglei II would disrupt human activities and animal husbandry in the southern Sudan, but add little value beyond that provided by Jonglei I.

Uganda currently has no plans for storage at Lake Albert, and even if it were to develop such plans it would encounter great difficulties in capturing the Congo's attention, much less interest, in what would have to be a bilateral, or, with the PJTC, a trilateral accord. Uganda's active sponsorship of dissident Congolese forces hostile to Laurent Kabila, and now to his son, make any deal with Kinshasa all the more unlikely.

CONCLUSION

We have now reviewed the water-related interests and strategies of the four principal riparians in the Nile basin. The Sudan is formally wedded to Egypt through the 1959 agreement, and that wedding has been reiterated by Sudanese officials in several public utterances since then. The Sudan's future interests, however, may well lie in some form of collaboration with Ethiopia on use of the Blue Nile. Egypt and Uganda are formally linked through the 1949 agreement allowing the construction and operation of the Owen Falls Dam. Many responsible Ugandans are sympathetic to any moves that they see as thwarting Egypt's drive to dominate the Nile basin, but the simple fact is that Uganda's interests in power generation can best be met by following the regime set down in the 1949 agreement. There would be little to gain and much to risk for Uganda to align itself with Ethiopia in challenging Egypt's assertion of its acquired rights.

In fact, Egypt can probably count on Uganda to protect its rights vis-à-vis Kenya, Tanzania, and Rwanda, were any of those Lake Victoria riparians to undertake projects that would significantly reduce outflows of tributaries into Lake Victoria, or, as in the Smith Sound project, draw significant amounts of water from the lake itself. Uganda would suffer first and foremost from any diminished discharge from Lake Victoria into the Victoria Nile. Sustained abstractions coupled with a lower lake level would put in jeopardy Uganda's plans for increased hydropower generation. Egypt would suffer relatively little, as abstractions in the Lake Victoria basin do not translate directly and proportionately into reduced flows out of the Sudd swamps and into the White Nile.

Were Jonglei I to be completed (possible) and were Jonglei II to follow (highly unlikely), then Egypt would probably seek a major change in the status quo, one altering the seasonal rhythm of discharge from Lake Albert and from the Albert Nile. This change would provide no direct benefits for Uganda, nor for the Congo, which shares Lake Albert, and thus both countries would require meaningful compensation from Egypt to make it worth their while.

Conclusion Lessons Learned?

The premise of this book is that cooperation in the use of transboundary resources is desirable and will tend to enhance the welfare of the greatest number of those who have access to or live from the resource. Enhanced welfare is, however, for the concerned policy-makers and potential beneficiaries, a hypothetical construct. For the rare few who try to quantify future benefits, there are the far more numerous constituents who feel comfortable with the status quo. Crisis in the quantity or quality of supply may drive users toward cooperation or, alternatively, to conflict. It is a sine qua non of cooperation that there be a consensus that the status quo is neither desirable nor viable. In the Nile basin there is not yet a crisis. There is no consensus that the status quo is nonviable, and at least two countries, Egypt and Uganda, are fairly firmly wedded to the status quo.

My purpose is not to lament this fact. The welfare of the inhabitants of the Nile basin is affected by many factors, foremost among them flawed governance and economic mismanagement. It cannot convincingly be argued that the absence of a cooperative regime in the basin significantly and negatively affects the welfare of the average inhabitant.

It is rather my purpose to use the Nile basin as a lens through which to view and isolate the dynamics that either impede or work toward cooperative understandings. The exercise has been designed to test some basic propositions on the dynamics of cooperation, on the processes by which regimes emerge or do not emerge, and on institutional solutions to collective action problems.

To revert once again to Lichbach's typology of frameworks for solving collective action problems (1996, and Chapter 1), a community of riparians does not exist in the Nile basin. There are no accepted norms of group behavior that could shame riparians into upholding group action. Market mechanisms are also very weak, but have been used on occasion (British-Ethiopian payments in the early decades of the twentieth century; Sudanese Eritrean payments at the same time; Egyptian compensatory payments to Uganda). Transboundary water markets do not yet exist (although one did exist on the Gash, shared by the Sudan and Eritrea), but it is desirable to promote them.

The main frameworks that can promote and sustain cooperation are contract and hierarchy. The catalyst to cooperation will be third-party entrepreneurs located in the donor community. The principal players are nation-states pursuing strategic and national interests. Any common action between or among them will be based on contractual agreements and understandings. It should be in the future, as it has been in the past, possible to understand most interactions among the players in rational actor terms.

It also is the case that two riparians are considerably more powerful than the other eight: Egypt and Ethiopia. Egypt, however, is by far the most powerful riparian. While its ability to project its strength may be waning, it still has a formidable veto power. It has been successful in imposing the status quo for four decades, and it will surely shape any change in the status quo. It cannot dictate terms, but no riparian, including Ethiopia, will seek, let alone welcome, confrontation with Egypt when its well-known national interests are at stake.

This apprehension is all the more paralyzing in that many riparians are indifferent to cooperation, unaware of the benefits that might flow from cooperation, or absorbed with issues of much higher national priority. We normally see defection as the threat to collective action, but defection assumes that the free-rider sees the benefits of the action and wants to share them at no cost to itself. The awareness of the benefits and thus the temptation of defection have not yet fully emerged in the basin.

The riparians can be grouped in three categories. The categories are not static and their composition may change over time.

Status Quo	*New Regime*	*Indifferent*
Egypt	Ethiopia	Kenya
Uganda	Sudan	Tanzania
	Eritrea	Congo
		Rwanda
		Burundi

I have placed Sudan among the riparians seeking a new regime mainly because of the unhappiness many Sudanese experts express with respect to the 1959 agreement. But it should be stressed that on Nile issues, the Sudan has seldom broken ranks with Egypt. Nonetheless, in the late 1950s it essentially repudiated the 1929 agreement with Egypt, and in the 1990s it has begun raising the height of the Roseires Dam without Egyptian approval.

Tanzania might become a spoiler of sorts if it were to pursue the Smith Sound project (see Chapter 7). The other riparians, above all Uganda, but also Egypt and the Sudan, would then not allow Tanzania the luxury of indifference.

Egypt could alter an element in the status quo were it to pursue excavation of a second Jonglei canal or seek the doubling in capacity of Jonglei I. Either option would entail complicated negotiations with the Congo and Uganda and a new regime in the upper White Nile and Victoria basins (see Chapter 7). Otherwise Egypt's main goal is to seek basin-wide endorsement of the status quo.

The Nile basin is essentially bipolar. It is dominated by a low-key but persistent confrontation between Ethiopia and Egypt that has been little affected by political regime changes in either country. The bipolarity existed even in the precolonial and colonial eras, but in both earlier eras, more than today, the great powers interacted with and manipulated a diverse array of local actors and interests. It was during the century or so preceding the Cold War that the three-level game analyzed in Chapter 3 was at its height.

Since World War II and the advent of the Cold War, the primary actors have been the states of the basin, and they have behaved as unitary actors. There has yet to be any significant impact of domestic interests and public opinion upon the narrow policy-making groups in each country who formulate national strategies toward issues affecting the basin. Only with respect to the interminable civil war in the southern Sudan and the thirty-year struggle that ended in Eritrean independence have outside powers found nonstate actors with which to interact. Hence, it is crucial in order to assess the prospects for cooperation in the basin to understand the strategic interests, defined by national policy-making elites, of each country. This was done for Ethiopia (Chapters 4 and 5), the Sudan (Chapter 6), and Uganda and the Lake Victoria riparians (Chapter 7).

In the Egypto-Ethiopian confrontation, the Egyptians have tried to build an agenda around *process* issues whereas Ethiopia has stressed *principles*. Egypt, as the main beneficiary of the status quo, proposes processes by which the needs of all riparians can be identified and the expertise and resources mustered to meet them. The focus on process suggests problems that are not zero-sum and in which all parties can meet their needs without calling into question the two-state regime constructed on the basis of the 1959 agreement.

For Egypt, cooperation means sharing in basin-wide efforts to enhance the flow of the Nile and its tributaries (enlarging the pie) so that any needs of the upper basin riparians can be met out of the incremental gains. In this view the 1959 allocation, the status quo, is protected by the principle of acquired rights while equitable use governs the allocation of incremental gains from en-hanced flow. Would Egypt, vis-à-vis the upper basin riparians, follow the same allocational logic enshrined in the 1959 agreement by which the Sudan received two-thirds of the incremental gains? (see Chapter 3).

By contrast, Ethiopia sees the struggle as zero-sum, that change can come only through the dismantling of the 1959 regime, and that to seek so bold an outcome of necessity requires an argument based on principle, specifically on equity embodied in a users' code.

In sum, Egypt argues that unless water supply in all its forms (rainfall, groundwater, nonbasin water resources) is harnessed and expanded, then any change in the status quo will cause it harm. To avoid that, all riparians should cooperate in finding the technical and financial means to enhance supply as well as to seek development that does not depend on Nile water. Ethiopia sees such arguments as formulae for stalling and avoiding the real issue, which is reapportionment based on equitable use for *all* the water in the Nile, not merely enhanced flows over and above the 84 bcm used in the 1959 agree-ment.

The most concerned riparians can be grouped schematically according to their espousal of or claims to protection under doctrines of appreciable harm or of equity in use (see Chapter 1). The countries I have listed above as indifferent may have firm views on the two principles but insufficient interest in the re-source itself to put much political capital on the line in their defense.

	Equity		**Harm**
Poverty	*Dependency*	*Potential*	
Ethiopia	Egypt	Sudan	Egypt
South Sudan	Uganda		Sudan
			Uganda

If equity is partially defined as protecting the supply of water to poor populations that rely heavily upon it for their livelihood, then Ethiopia and the southern Sudan have strong equity claims. Similarly, if national dependency upon the basin's resources is a major element in claims based on equity, then Egypt, with no source of water other than the Nile, and Uganda, whose territory lies entirely in the Nile watershed, likewise have strong claims based on equity. Finally, if potential for water-based development is a valid equity claim, then the Sudan has vastly more potential than any other riparian. It is for that reason that it will have to seek a change in the status quo.

Egypt, however, bases its policies on defending against any harm to its existing uses of the Nile, which, it claims, grants it senior rights. Likewise for the Sudan the apportionment rendered under the terms of the 1929 and 1959 agreements accorded it rights to water that are second only to Egypt's in seniority. Uganda also could suffer appreciable harm were the Lake Victoria states to increase their use of the normal discharge of rivers into Lake Victoria or to pursue projects to take water from the lake for their own use. On that score Uganda is a defender of the status quo in the Lake Victoria basin and, inadvertently, a watchdog for Egypt over one part of the Nile catchment.

The fact that some riparians can be listed simultaneously as beneficiaries of doctrines of harm and of equitable use underlines an important facet of negotiating for new regimes or frameworks of cooperation. Ignorance, I posited in Chapter 2, can be beneficial in the initial design of regimes. That is, if there are important gaps in the knowledge the parties have of one another and of the resource itself, then it is more likely that a "just" regime will be designed, one that protects all ignorant parties against the unforeseen. There are in the Nile basin only two proxies for ignorance, but they are substantial. One is inconsistent incentives (see Chapter 2). Some riparians, like the Sudan, are both upstream and downstream, possible beneficiaries of Harmon-like doctrines of absolute territorial sovereignty over the resource and possible victims. To protect oneself against this inconsistency may make for a more just regime. The second area of ignorance is the relative difficulty in knowing the evolution of the natural regime of the resource itself. What kind of secular trends in temperature, rainfall, erosion, and sedimentation, can one anticipate? Any formal regime will want to protect its members against changes in the quantity or variability of the supply of the managed resource.

Unfortunately there is very little incentive inconsistency for two of the principal riparians in the Nile basin. Egypt is a pure downstream state, receiving all its surface water from outside its borders. It cannot trap itself by single-minded adherence to the principle of appreciable harm. Similarly, Ethiopia receives no water from outside its borders. It is a pure upstream state. The Harmon Doctrine is politically incorrect, but were it more acceptable in international circles, Ethiopia would almost certainly espouse it. All other riparians have inconsistent incentives, but only two, Uganda and the Sudan, have major stakes in the Nile basin.

While it can be argued that Egypt's stake in the Nile is diminishing over time, it is indisputable that the stakes of Ethiopia, the Sudan, and Uganda are growing. Egypt's economy is increasingly dominated by the nonagricultural sectors, and its need for power from the Aswan High Dam is becoming less and less significant. A plausible case can be made that Egypt's economy would not much suffer from some diminution in its share of the Nile's discharge. The fact that it is, instead, pursuing projects, like the New Valley scheme, that will create new demand for water, is profoundly discouraging to the Ethiopians and, perhaps, to the Sudanese. In other words, Egypt's definition of its national interests is rooted firmly in the objectives of the 1950s while its economy moves rapidly into the twenty-first century.

Ethiopia is still, and for at least a generation will remain, a peasant society whose economy will be dependent upon the performance of the agricultural sector. That sector, in turn, is dependent upon highly variable and unpredictable rainfall. Famine has struck Ethiopia roughly every ten to fifteen years (there is absolutely no science in such figures), and in two instances contributed directly to the collapse of governments and regimes. At the turn of the century famine had struck again, almost "on the decade." With a rapidly growing population and a degraded rural environment, it is quite plausible to argue that Ethiopia must, even at great cost, introduce some predictability into the agricultural sector. In Chapters 4 and 5, I explored how that might be done. There are neither simple nor uniformly efficacious remedies, but harnessing Nile tributaries for irrigation and power generation will surely be part of the solution.

Ethiopia is pushing its cause, inconsistently and clumsily, now. The Sudan may do so in a decade, perhaps with greater skill. The Sudan could be a major agricultural producer and exporter. It has all the ingredients: good lands that can be brought relatively easily under irrigation and the plow, abundant surface water resources, and no obvious alternatives to driving the economy.[1] Ethiopia's most ambitious potential demand for Nile water would not exceed 4–5 bcm. The Sudan might one day claim 30 bcm or more, nearly double what it was allotted under the 1959 agreement with Egypt. Ethiopia could rattle the status quo. The Sudan could smash it to smithereens.

Uganda's agricultural sector can develop with existing surface water resources and some supplementary irrigation. The current government has focused on industry and other nonagricultural sectors (tourism, services) to carry the economy forward. In that sense, hydropower availability is crucial to supply domestic urban and industrial demand and to export to neighbors. The regime of water release from Lake Victoria set up in 1949 with the agreement between Egypt and Uganda (then under British control) that allowed the construction of the Owen Falls Dam at Jinja still operates in Uganda's favor. The agreement was designed to protect Egypt's needs, before the construction of the Aswan High Dam, for summer or "timely" water. Even though Egypt no longer needs timely water now that it has the Aswan High Dam reservoir, Uganda needs the 1949 regime.

If the Sudan asserts its claims in the future as outlined above, it would almost surely do so on the Blue Nile tributaries (see Chapter 6). Uganda would not be directly affected. At the same time, the Sudan and Ethiopia might be drawn into alliance for the development of the Blue Nile and into integrated infrastructure spanning their common border. That, I have argued, coupled with a doubling of Sudanese demand for Nile water is Egypt's worst nightmare. It need not be, if Egyptian policy-makers looked forward to the role agriculture is likely to play in Egypt's economy, rather than backward to a now-sacrosanct status quo defined by the 1959 agreement. The fundamental question for Egypt, the major power in the basin, is whether to do everything in its power to avoid the nightmare or to develop plans to cope with it.

Egypt, at least since the days when Boutros Boutros-Ghali was Minister of State for Foreign Affairs, has sought a basin-wide regime that would embody the acquiescence of all the other riparians to the basic allocation laid out in 1959. Ethiopia has sought a basin-wide regime based on a new framework or users' code that would undo that allocation. At this point in time, were the contest between only Egypt and Ethiopia, Egyptian policy-makers would have little to trouble their sleep. The only element that might arouse their concern is the donor community itself. The World Bank and the UNDP, through the Nile Basin Initiative, have tried to reconcile Ethiopia's concern for a framework to promote equitable use of all Nile water available with Egypt's concern to emphasize multisector and multi-good cooperation, coupled with enhanced use of all forms of surface water supply in the upper basin.

The World Bank has shown its readiness to use resources to bring about some sort of regime in the basin, and such a regime, as evidenced in the Nile Basin Strategic Action Program, would hew more closely to the principle of equitable use than to appreciable harm, and it would foster some form of reallocation. The NBI does not step beyond one of the declared functions of the Tecconile Agreement, December 8, 1992, which Egypt both sponsored and signed: "To assist participating countries in the determination of the equitable entitlement of each riparian to the use of the Nile waters" (as cited in Ntambirewki and Mujaju, 1993). The NBI sets as its goal "to achieve sustainable socioeconomic development through the *equitable utilization* of, and benefit from, the common Nile Basin water resources."[2]

One of the lessons of this book is that the search for comprehensive agreements or regimes, without several intervening steps, is problematic. There are too many players and their interest in the resource and the benefits they are likely to draw from basin-wide cooperation are highly asymmetrical. It follows that policy attention in those with the least to gain is, virtually by definition, focused elsewhere. Very clear gains, probably in the form of monetary compensation, must be offered to the indifferent in order to win them over. The status quo does them no harm. A new regime would be attractive only if the benefits significantly outweighed the costs. Third parties, such as the World Bank and the African Development Bank, could provide some of the compensatory benefits. That might launch a regime, but would it sustain it?

I have argued that it may be more useful and productive over the long run to lay the foundations of a regime through two intermediate steps. The first is to begin at home (see Chapter 1). By this I mean undertaking those policy initiatives to encourage efficient use of water and to protect water-sensitive environments that any riparian would or should want to carry out regardless of its relations with other riparians. In this respect Uganda's 1995 National Environment Statute is exemplary (see *Uganda Gazette,* 1995). These steps could involve levying charges for water or types of water use that would directly or indirectly reflect water's scarcity value and enter that value into all financial and economic return calculations.

In this manner each riparian would independently begin the process of meeting best practice standards in the use of water. Supply would be governed, after accounting for the basic needs of all inhabitants, by market measures, while quality would be governed by concerns for hygiene, public health, and environmental protection. When and if the day comes that all the riparians, or some subset of them, wish to move toward a basin-wide accord, they will be able to compare best practice benchmarks and agree upon standards in water use that will facilitate discussions of real needs and of quality.

The second *démarche* is to seek cooperation at the sub-basin level (see Chapter 2). In addition to the 1959 agreement, which essentially involves the lower basin at the expense of the rest, at least six sub-basin accords could be envisaged.

1. The Lake Victoria states are already loosely grouped since 1996 in order to combat the water hyacinth and to reduce pollution (see Chapter 7). This could be the building block to an accord with wider scope including all projects involving abstraction from, or runoff into the lake, lake transportation, hydropower sharing, disease control, and the like.

2. The Kagera Basin Organization already exists, although it is moribund at the moment (see Chapter 7). It could be revived along the lines originally laid down in the mid-1970s.

3. Uganda-Sudan would manage the Albert Nile and the Bahr al-Jebel. This would come into play for joint management of the Sudd swamps, perhaps coupled with power generation on these portions of the Nile. It would almost surely be contingent on some sort of settlement to the civil war in the southern Sudan and its counterpart, the Lord's Resistance Army, in northern Uganda (Chapters 6 and 7).

4. Uganda-Congo-Sudan-Egypt would manage the entire White Nile. This would come about if Jonglei II were to be pressed by Egypt and were Lake Albert to be designated as the primary storage site for the water, to be released during the dry season through the two parallel canals (or one large canal).

5. Sudan-Ethiopia would develop the Blue Nile and perhaps the Teccaze-Atbara and the Baro-Akabo. In my view, and in the long run, the logic of such sub-basin development is compelling. It would involve water storage and power generation in Ethiopia with irrigation schemes on both sides of the border and an integrated power grid (Chapter 6).

6. Eritrea-Sudan would manage the Gash and Baraka rivers. These are fairly minor seasonal flows, but very important to the regional economies of the northeast Sudan and the southern portion of Eritrea.

Sub-basin accords could one day "federate" into a basin-wide accord. The impetus in that direction would be no stronger or weaker than the impetus to link international free-trade zones. Some of the sub-basin accords already exist, and others appear to be probable. Certain countries, such as Uganda and Sudan, are likely to be party to more than one because of their strategic location. Still, the interlinking of sub-basin units in a comprehensive accord is more possibility than probability.

At whatever level, allocation will initially involve a somewhat arbitrary assignation of "rights." Because it will be arbitrary there must be a strong sense among the parties that the status quo is not viable or that the benefits of cooperation, even if asymmetrically distributed, are of such magnitude as to drive the project forward. We should not put too fine a point on this: it will be rug selling, splitting differences, and swallowing hard. In one respect the 1959 agreement is exemplary. It used mixed criteria in determining different parts of the allocation (see Chapter 3), and, before the principles had become the subject of wide international debate, saluted both appreciable harm *and* equitable use.[3] The agreement partially failed because it reduced flexibility in two ways. It eschewed proportionality in favor of allocating fixed quantities (such an allocation means any change becomes zero-sum), and it did not allow for trading "rights." I have stressed that an initial, arbitrary assignation will be sustainable only if the parties to it do not feel trapped by its terms. Even if it be at the margins, the possibility of acquiring more water from other riparians must be formally recognized and the infrastructure put in place to make it a reality.

A key interest group may be formed in all stages of these processes: in the development and implementation of domestic water and environmental policies, in the negotiation and design of sub-basin accords, and in the steps toward basin-wide framework agreements or formal and binding accords. It is the group of legal, economic, and engineering professionals who will have to carry out the drafting, debating, and designing of the accords and attendant policies. They will share a conceptual language and probably a set of priorities if not values. Such groups have been called by Peter Haas (among others) "epistemic communities," and they may be crucial to sustaining a new regime (see Chapter 1).

Finally, there is likely to be shared infrastructure in the form of water control structures, water delivery infrastructure, power grids, road and railroad links. Logically, the countries that invest heavily in and jointly develop this infrastructure will not want to close it down, let alone destroy it in hostilities. That logic does not always work, as has been shown in the closing down of petroleum pipelines shared by Iraq and Syria and by Iraq and Turkey. By contrast the construction and successful operation of natural gas pipelines from the Soviet

Union to Western Europe during the Cold War demonstrates the logic somewhat better.

The riparians of the Nile basin have the luxury of pursuing these steps sequentially, because the wolf of unsustainable use is not yet at their door. Upstream states in any sub-basin may reduce supply or lower the quality of the resource, thereby undermining the efficacy of the domestic measures to manage supply and quality. However, in the Nile basin only the Sudan, at present, could change either dimension to inconvenience another riparian, Egypt. The basic fact is that some progress has been made at all three levels. Uganda is not alone in drafting ambitious national legislation and policies (implementation, of course, is another matter). Sub-basin accords already exist, and in the case of Egypt and the Sudan function quite well. Since the launching of Hydromet and onward through the Undugu group, the Tecconile grouping, and the Nile Basin Initiative, basin-wide debates have been carried forward regularly, if not smoothly. Elinor Ostrom's summary on institutional change is apposite: "The investment in institutional change was not made in a single step. Rather, the process of institutional change in all basins involved many small steps that had low initial costs. . . . Each institutional change transformed the structure of incentives within which future strategic decisions would be made" (Ostrom, 1991:137).

The initial costs of noncooperation in the Nile basin have been significant. The costs of infrastructure in Egypt, the Sudan, Ethiopia, and Uganda have been substantial. But we are, after all, talking about the Nile. Egypt's initial commitment to the New Valley project will undoubtedly set back the prospects for further steps toward basin-wide understandings, but it need not kill them. Debating this project with Egypt may establish that downstream states, contrary to geographical logic, can cause appreciable harm to upstream states by preempting their options, in short, by foreclosing the future.

While in juridical terms all ten riparians in the Nile basin count equally, in terms of real stakes in the basin there are only four concerned actors, with a fifth unrecognized by any of the others. Egypt, the Sudan, Ethiopia, and Uganda, were they ever to come to terms, could by themselves provide a new regime for the Nile. Note that they are in no way a community; they do not share norms nor can they exert social constraints on one another. Whatever compact they conclude will be on the basis of the rational pursuit of their national interests. This allows us to interpret and to anticipate their moves in terms of maximizing benefits and of avoiding harm.

The fifth player is whatever group or groups that speaks for the southern Sudan (see Chapter 6). With potential control over the Sudd swamps and the discharge into the White Nile, the southern Sudan and its peoples are vitally concerned parties to any regime. While Egypt and the Sudanese government would like to see the southern conflict resolved without jeopardizing the 1959 status quo, Uganda and Ethiopia (and, for a time, Eritrea) use the conflict to gain leverage over the Sudan and over Egypt. It is not clear under what circumstances and on what terms either Uganda or Ethiopia would like to see the civil war in the southern Sudan concluded.

That said, one can envisage two broad solutions to the cooperative challenge in the Nile basin. There could be a quasi-hegemonic solution in which Egypt imposes a regime upon the Sudan and Uganda. It is not likely, however, that Egyptian leverage over Ethiopia would ever be sufficient to force it to adhere to such a regime. But whether a three- or four-party understanding, it would not be voluntary.

A voluntary solution would involve Egypt acquiescing to a new regime, one in which it would concede some of its acquired rights under the 1959 agreement. That would come about because Egypt redefined its national interests to emphasize trade with its Nile basin neighbors and to reassess its need for water in light of the profound transformations in its economic life.

Such a solution would one day have to face the challenge of a Sudanese attempt to bring about a radically changed regime that would see Egypt's claims to water reduced by as much as half. The Sudan would not mount this challenge without at least the tacit support of Ethiopia.

How much significance should we attach to the fact that the few existing formal accords among Nile riparians have been concluded under hegemonic auspices? The 1929 agreement between the Sudan and Egypt was imposed by Great Britain, the 1949 agreement between Egypt and Uganda was heavily influenced by Great Britain, and the 1959 agreement was negotiated under a Soviet umbrella.

By contrast, examples that may presage future accords are the Kagera Basin agreement, pushed by Belgium but funded through the European Community, and the Lake Victoria grouping brought about almost entirely by the coaxing of the World Bank. Only if the United States and the donor community committed themselves unquestioningly to the defense of Egypt's interests in the basin as currently defined would there be a quasi-hegemonic solution to the collective action problem, and it would take the form of regime maintenance and the consolidation of the status quo. Egypt's interests will not be lightly dismissed by any third party, but most third parties are well aware that Egypt is now economically strong enough to weather changes in the status quo. The other riparians may not be economically strong enough to live with the status quo, and that fact may dominate the future.

Notes

1. The Common Market, the European Community, and the European Union are designations that apply to the same entity in Western Europe at different points in time. Other grain exporters of significance in the early 1970s were Brazil (soy bean and soy products) and Thailand (rice).
2. The Dutch Disease, so named because of the effects of natural gas exports on the Dutch economy, is the result of the influx of natural resource export earnings on the relative prices between traded and nontraded goods in favor of the latter. As wages and investment rise in the nontraded (mainly services) sector, the traded-goods sector (mainly agriculture and manufacturing) tends to stagnate. Petroleum exports have had that effect on most exporting countries. In the Middle East, oil-importing countries sent millions of laborers to the oil-rich but underpopulated exporting countries. Worker remittances into these economies had the same effect on traded/nontraded-goods sectors as did petroleum exports in the economies of the oil rich (see Gelb et al., 1988).
3. Some legislators in developed countries that export agricultural commodities oppose credits and sales to developing country projects that may compete with these exports.
4. For a full inventory of international watercourses, see UN (1978) and the excellent updating in Wolf et al. (1999).

CHAPTER 1. COLLECTIVE ACTION AND THE SEARCH FOR A REGIME

1. Elinor Ostrom extends this analysis to renewable resources in general where the resource can be conceived of as a stock and the renewable "use units" as flows (Ostrom, 1991:30).

2. It is use, and not consumption, which is at issue. Water that is used does not disappear. It is returned to the hydrological cycle through the evapo-transpiration of plants, surface evaporation of reservoirs, lakes, and seas, and even through decomposition of organisms that have absorbed it.

3. The most prominent sub-basin organization is the Mekong Commission, grouping the riparians of the lower basin. Some possible sub-basin groupings would be Kyrgyzstan and Tadjikistan versus Kazakhstan, Turkmenistan, and Uzbekistan; or Sudan and Ethiopia versus Egypt (see Chapters 2 and 6).

4. Karl Wittfogel (1957) analyzed the dynamics of the great, ancient river civilizations of the Yangtze, the Indus, the Tigris-Euphrates, and the Nile.

5. In 1957 Canada rejected U.S. claims to uses of the Waterton and Belly rivers, arising in Montana and flowing into Alberta in Canada. The Canadians claimed prior appropriation and noted that Montana had not used the water. Canada argued that "past use conserved the right for future use." All that could be negotiated was apportionment of the flow remaining in excess of Canada's use (McDougall, 1971:271). This is exactly Egypt's position toward all Nile riparians save the Sudan, with which it signed an agreement in 1959.

6. Huffaker et al. (2000) note that inefficient use of water in one part of a watershed can provide benefits for users further downstream. Efficiency may thus cause harm. It is the state's role as trustee, the authors argue, to assess the welfare consequences of rival uses (pp. 270–71).

7. Article 7 enjoins riparians from causing significant harm to other riparians, and Article 12 instructs riparians undertaking projects that might cause harm to other riparians to notify them of their intentions and to supply them with the technical data used in project design and justification. The text of the 1997 convention appears in Salman and Boisson de Chazournes (1998:178–92).

8. The 1997 ILC text does not deal with groundwater or aquifers. They present peculiar problems. On the one hand, groundwater is hydrologically part of surface watercourses. If the basin is the proper unit of analysis, then its groundwater must be part of the equation. However, transboundary groundwater can be used simultaneously by riparians and in that sense is like petroleum. Caponera and Alhéritière (1978) therefore stress the community of interest principle in regulating groundwater use and reject prior appropriation. If the latter principle were applied, it would trigger a race to exploit the slowly recharging resource (p. 616).

9. Equitable use was given a major boost when Hungary and Slovakia referred their dispute over Slovakia's diversion of the Danube at Gabcikovo and Nagymaros to the International Court of Justice. The court rendered its judgment in September 1997 and in its ruling endorsed the principle that "the community of interests in a navigable river becomes the basis of a common legal right, the essential features of which are the perfect equality of all riparian states in the uses of the whole course of the river." See Sands (1998:105).

10. So too Székely (1990), arguing from the standpoint of Mexico; he insists that apprecia-

ble harm must be stringently defined, and that the notion of cumulative harm be included in the definition.

CHAPTER 2. NEGOTIATING REGIMES

1. See Bilen, 1997; Chomchai, 1995, Crow, 1995; Crow and Singh, 2000; Godana, 1985; Kolars and Mitchell, 1991; Lowi, 1993; Micklin, 1991; O'Hara, 2000; Salman and Boisson de Chazournes, 1998.

2. Supalla (2000:261) reviews in game theoretic terms the formation of a regime in the Platte Basin, grouping Colorado, Wyoming, and Nebraska. He observes, "No interest group was willing to agree to the facts that did not support their case." He concludes with two lessons. "The first was the futility of relying on science as the basis for a solution and the second was the importance of having all parties share a mutual interest in reaching an agreement."

3. Lilienthal cites the first generation, and claims credit for the TVA in inspiring authorities in Australia, the Philippines, Iran, Chile, and India, as well as transnational bodies on the Jordan, the Nile, the Niger, the Rhone, and elsewhere.

4. Oran Young (1994:72) shows, however, with respect to Arctic accords, that bilateral agreements are no more effective than multilateral agreements.

5. Boggess et al. (1993) show comparative returns to water use in the United States for 1983. An acre-foot in agriculture returned $9 to $103; domestic use $19 to $322, industrial use $0 to $160, and recreational use $3 to $17. Water cannot be allocated infinitely to the use with highest return because of diminishing returns. Industry, for example, could never absorb more than a fraction of the water used in agriculture.

6. Because Turkey did not secure the acquiescence of the downstream states to the several projects already under way or completed, the World Bank could not participate in funding them. Turkey's economy is strong enough, however, to allow it to borrow in commercial markets. The large international engineering firms and suppliers of capital goods are happy to advance suppliers' credits for Turkey's needs. Iraq is urging the Arab League of States to boycott those firms.

7. At the UN Water Conference of 1977, held in Mar del Plata, Argentina, upstream and downstream countries represented there cast off Cold War positions to regroup according to their geographic incentives. In working groups, policy debates, and in the plenary, Iraq, Mexico, Bangladesh, Argentina, Rumania, Egypt, and others collaborated in defending downstream interests, while Turkey, India, the USSR, Guinea, Ethiopia, Peru, Ecuador, and others did the same for the upstreamers (see Waterbury, 1977:21).

8. Stephen McCaffrey has proclaimed "absolute territorial sovereignty" dead (McCaffrey, 1998:23). It is no longer politically correct to invoke it, or the Harmon Doctrine, by name, but, as the examples cited show, its spirit is very much alive.

CHAPTER 3. THE THREE-LEVEL GAME IN THE NILE BASIN

1. This is merely figurative. In fact most of the water in the Abbay-Blue Nile is delivered by tributaries that join the river *after* it exits Lake Tana.

2. International law wrestles with the issue of the extent of a successor state's obligation to abide by treaties signed by its predecessor(s), often with authorities that no longer exist. In particular, former colonies have frequently threatened to repudiate obligations undertaken in their name by colonial authorities. Because these obligations potentially bind their independent neighbors, the threat of repudiation has not often been put into practice.

There are two polar positions on this issue: the so-called clean-slate principle, which posits that with independence all colonial treaty obligations automatically lapse, as opposed to the principle that successor states are completely bound by obligations undertaken by their predecessor states, colonial or otherwise. Fisseha (1981:187–89) argues that when the Sudan, in 1958, repudiated the 1929 agreement binding it and Egypt to an apportionment of the Nile, it simultaneously repudiated the 1902 treaty with Ethiopia. In 1958, Sudan argued that the 1929 agreement was technically and economically obsolete. Ethiopia, so argues Fisseha, could repudiate the 1902 treaty on the same grounds.

3. Zewde (1976:41) notes that in 1924 Ras Tefari approached the British Sudan with an offer to undertake the construction of the regulating dam and power station on Lake Tana and to sell the stored water to the Sudan. Zewde does not say how this proposal was received, but Ras Tefari may have sensed that British colonial policy was not monolithic and that there was a chance to split off the Sudan from Egypt. It did not succeed, but it has always remained a valid option for Ethiopia (see Chapter 6).

4. I have relied heavily in my discussion of the Tana concession on the excellent article by James McCann (1981).

5. Until the construction of the Aswan High Dam, all storage sites on the Nile were seasonal in nature; that is, they could capture part of the annual flood, releasing the bulk downstream and ultimately to the Mediterranean. That meant that a series of low years would leave Egypt exposed to severe water shortages in agriculture. The Aswan High Dam was designed to impound up to two consecutive floods (based on established mean discharge) and thus hold in storage enough water to protect normal levels of agricultural production for two years or more.

6. Details on this study are presented in Chapter 5.

7. I owe this interpretation in part to Kifle Wadajo, interview, October 18, 1995, Addis Ababa.

8. These rumors have been a constant in Nile politics, and I suspect they have been allowed to run by both Ethiopia and Israel. I have found no concrete evidence that any project was discussed in more than hypothetical terms, but the suspicion alone could keep the adversaries of both countries guessing.

9. Even though Ethiopia hosted the fifth Tecconile conference in early 1997, it still remained in observer status in that grouping.

10. I explored this theme earlier in Waterbury (1997).

11. As was pointed out later, such seepage would probably return to the system further downstream and thus would not constitute a pure loss.

12. The full title is Agreement Between the Republic of the Sudan and the United Arab Republic for the Full Utilization of the Nile Waters. It was signed in Cairo on November 8, 1959. The full text is in UN, 1963.

13. But not entirely; the agreement of December 8, 1992, establishing the Tecconile group of Nile riparians, declares as one of its purposes "To assist the participating countries in the determination of the equitable entitlement of each riparian to the use of Nile waters" (as cited in Ntambirewki and Mujaju, 1993).

14. I am grateful to Nicholas Hopkins for sending me this article. Badr's article is written in much the same spirit as that of Hilmy (1978), although the passage of seventeen years between the two makes Badr's all the more unusual.

15. The World Bank has created a world water atlas of extraordinary detail, based on satellite imagery, and available on the internet. If water and its location/distribution are strategic matters, one's grandmother can now access the family jewels. The web site is *www.iwmi.org.* See "Secrets of Nations' Water Security Exposed on the Internet," *al-Hayat,* December 9, 1998, p. 14 (in Arabic).

16. Hydromet, by any measure, was not a notable success. Much of the infrastructure (gauges and the like) was destroyed in the region's various civil wars. Political upheaval and periodic purges thwarted the buidling up of technocratic expertise. The mathematical model was flawed and soon became obsolete, and the quality of the data is suspect. A harsh critique is to be found in UNDP (1989).

17. It is said that the Minister of Irrigation and Public Works of Egypt, Abdelhadi Radi, approved of D3 with great reluctance.

18. One can find this approach amply reflected in a 1989 study carried out by the UNDP on behalf of the Undugu Group. The report (UNDP, 1989) takes as given the shares of Egypt and the Sudan under the 1959 agreement and concentrates on conservation projects that can enhance flows.

19. The Nile Basin Initiative has its own web site: www.*nilebasin.org.*

20. I have relied on two unpublished documents on the NBI: "Policy Guidelines for the Nile River Basin Strategic Action Program," Nile Secretariat, and "Nile River Basin Cooperative Framework: Panel of Experts Draft Cooperative Framework, Revision 2," Entebbe, December 10, 1999. I have not listed the documents in the bibliography.

21. The section heading is inspired by Waterbury and Whittington (1998).

22. As noted in Chapters 1 and 2, this principle of upstream-downstream reciprocity is not unambiguously recognized in international law (see Krishna, 1998:40; Huffaker et al., 2000:266).

23. Although Ethiopia does not recognize the 1902 agreement between Ethiopia and Great Britain as binding, the Teccaze basin was not explicitly mentioned in that agreement.

24. It has been suggested that the World Bank could finance improvements and alterations to *existing* structures on a watercourse without seeking the acquiescence of other riparians. Krishna (1998:41), however, points out that any alterations can only be "minor" in nature.

25. There is a brief sketch of the stakes of the ten riparians in the Introduction. A fuller discussion is in Chapter 7.

26. The two exceptions to this statement, as noted above, were the Eritreans in their struggle against Ethiopia and the southern Sudanese in their struggle against the northern Sudan.

CHAPTER 4. FOOD SECURITY IN ETHIOPIA

1. The FAO study of 1986 and the UNDP/FAO study of 1988 are the two most scientific renderings of the argument Hoben challenges, but many Ethiopian analysts have also adopted it: see Wolde-Mariam (1972) and Haile-Mariam (1995).

2. Examining the historical record left by travelers over the past five hundred years, Richard Pankhurst finds that major areas of the highlands, especially in Tigray and around Gondar, were close to treeless except for the areas around churches which were traditionally situated in groves with good water supply. The seat of the empire often had to move to find new sources of fuel wood. The settling of Addis Ababa in the second half of the nineteenth century led to the denuding of the Entoto hills. It was only through the importation of the seeds of the fast-growing eucalyptus tree that an adequate fuel source for Addis Ababa was secured (see Pankhurst, 1990:275–76 and 1992).

3. *Ensete,* or so-called false banana, is a root crop and dietary staple for the densely inhabited south-central provinces.

4. It is important to note that on average of the 6–7 million tons of grains pulses, and oil seeds produced annually, only about 1 million were marketed. Coffee, by contrast, was and is mostly marketed and heavily taxed by the state.

5. The poor distribution and inadequate supply of chemical fertilizers, coupled with more extensive cultivation and reduced fallowing, have given rise to the spread of a weed, known as *striga,* that grows only in depleted soils. Its appearance is the equivalent of dead canaries in coalmines.

6. Seyoum et al. (1995) have devised a "correlates of famine" analysis for Ethiopia, the key variables of which are the "normal deviation of vegetation index" (see IGADD, 1991, for an application to Ethiopia), per capita maize production, and infant mortality. Scores on these indicators help identify the most at-risk families and localities. The authors found that the most at-risk families had *small* family size, were dependent on no more than two crops for subsistence and cash, had few or no animals, and had low productivity per unit of land and labor. Two medium-term coping mechanisms are to help cultivators diversify cash crops and, concomitantly, link remote areas to the nearest roads.

7. At the beginning of the twentieth century, the region between the Angareb and Teccaze rivers that flow into the Sudan from northwest Ethiopia was used to provision Eritrea in sorghum and maize. Highlanders came down during the rainy season to farm the area. The local pastoralists, the Hadendowa and Bani Amer, were not involved in cultivation; rather, West African migrants known as *fallata* and Galla slaves were used as labor. After World War I, this experiment came to an end (see McCann, 1990), but was revived after World War II. Private tractor farms were established on some 150,000 hectares in the Humera-Seteit region (Wolde-Mariam, 1972:116), and today these northern lowlands produce oil seed crops (especially sesame) that are marketed in the Sudan.

8. Kloos (1991:299) estimates that earthen microdams cost about $500 per hectare served, and, because of excessive sedimentation, their lifetime is short. Thus, if microdams on average serve an area of 300 hectares, each dam would cost about $150,000, the annual income of 15,000 "average" Ethiopians. In most instances local communities cannot mobilize that kind of money; only the donors and the central government can.

CHAPTER 5. THE IMPERFECT LOGIC OF BIG PROJECTS

1. The Awash River flows north toward Djibouti but ends, according to Ethiopia, in the salt lake, Abe, that straddles the border. Similarly, the Omo River flows south toward Kenya but empties into Lake Turkana (sometimes known as Rudolph) at or just north of the border. The remaining transboundary rivers are: the Wabe Shebele and the Juba, flowing into Somalia; the Akabo and Baro, flowing into the southern Sudan; the Abbay, Dinder, and Rahhad. which constitute the Blue Nile in the Sudan; and the Angareb and Teccaze, which join in the Sudan to form the Atbara.

2. Even in 1990 prices, the EVDSA-WAPCOS study estimated the cost of developing the 85,000 GWh at $19 billion, a sum that would choke a far stronger economy than Ethiopia's.

3. Sudan is pressing ahead with its development of its petroleum reserves in the Bentiu region in the south-central part of the country, and the Sudanese government may prefer to fill its power deficit through thermal generating facilities. I am not aware of any studies that compare the estimated costs of transboundary hydropower delivery from Ethiopia to those of thermal generation, or the forgone revenues Sudan would entail if it uses oil in its domestic economy rather than exporting it.

4. The 1986 study carried out by CESEN-Ansaldo and Finmecanica for the Ministry of Mines and Energy contains this statement: "It should be borne in mind, moreover, that high levels of solid transport, typical of the Tekeze Basin river course, could rapidly compromise reservoir availability. This is a factor of priority consideration in any more in-depth assessment of hydropower exploitation in this basin" (p. 43). The study itself makes no such estimates. The 1995 NRG study has this promise: "In Phase II existing sediment data will be verified and more sediment sampling will be conducted at selected stations" (chapter 5, p. 22).

5. By 1996, after thirty-two years of operation, Koka had accumulated 800 mcm of sediment or 44 percent of its total capacity (*The Monitor* [Ethiopia], April 18, 1996: 7–8).

6. Major Cheesman was British consul at Dangila, near the small tributary known as the "little Abbay" which flows into Lake Tana. His presence was testimony to British concerns. See his extraordinary accounts of his exploration of Tana and the Blue Nile (Abbay) watercourse over the period 1926–34 (Cheesman, 1968, and Chapter 3).

7. A preliminary study had been conducted by Tom Clarke of the Bureau of Reclamation in 1952, when both Egypt and Ethiopia were relatively friendly toward the United States. By 1958 that situation had changed radically.

8. This is typical of the gross misinformation often manifested in press treatment of Nile issues. Ahmad Sharif al-Din, writing in the Sudanese newspaper *al-Sudan al-Hadith* (October 21, 1995), claimed that the Finchaa-Amarti project stores 51 bcm, that it lies on the largest tributary to the Blue Nile with 83 percent of its total discharge, and that the dam and project caused Sudan's drought of 1984–85. I am grateful to Heather Sharkey for sending me this article.

9. In the drafting of the ILC Convention on nonnavigational uses of international watercourses, Ethiopia, along with Turkey and Rwanda, rejected a binding obligation to prior notification of works to be undertaken in the watercourse (see McCaffrey, 1998:41). It is

also the case that Ethiopia is under no treaty obligation with Egypt or the Sudan to notify them of works on the Teccaze.

10. I have used the per hectare water duty calculated in Transitional Government of
Ethiopia (1995: chap. 5, p. 60) for the Teccaze-Anghereb region, assuming a cropping in-
tensity of 1.5 to 2 per year.

11. See the excellent doctoral dissertation of Bahru Zewde (1976).

12. Some 400,000 refugees were moved in, but by 1995 only 70,000 remained. Most preferred to return to the highlands.

13. Although I have no systematic documentation, partial observations and limited studies
suggest that refugees in many parts of Africa have little choice but to be ecologically destructive. Often they "invade" areas of settled property rights (whether individual or
communal). It is difficult for them to rent or acquire land, and there may be little demand for their labor. What they can do is invade forest and savannah, gather wood and
other biomass for sale, use it to manufacture tools, utensils, and furniture, or turn it into
charcoal.

14. I am grateful to Steve Colombi-Brichieri for pointing this out to me; more generally I
would like to thank Gerard Chetboun, project manager for the TAMS-ULG survey and
master plan, and Solomon Kasahun, project manager in the Ministry of Water Resources, for their general observations on the project area. Neither Chetboun nor
Kasahun necessarily shares my understanding of the challenges involved in developing
this region.

CHAPTER 6. THE SUDAN: MASTER OF THE MIDDLE

1. Under the terms of the 1959 agreement with Egypt, the Sudan's total share in the Nile
flow is 18.5 bcm as measured at Aswan, or more like 20.5 bcm in the area upstream of the
confluence of the White and Blue Niles at Khartoum. The difference is the conveyance
loss (seepage and surface evaporation) incurred as the Nile flows north to Aswan.

2. In late August 1999 Sudan began to pump oil through a 1,600-kilometer pipeline from
Bentiu to Port Sudan (*al-Hayat,* August 29, 1999).

3. In the Sudan's Country Paper at the Fifth Nile 2002 Conference in 1997, the water needs
for the country to meet food security through irrigation were put at 32 bcm. That figure
does not include additional water needed to produce an exportable surplus.

4. For example, Sudan's former Minister of Irrigation, Yahya Abdel Mageed (1994), stressed
that the 1902 agreement (see Chapter 2) still binds Ethiopia, and that the 1929 agreement
between Egypt and the Sudan, while superseded by the 1959 Agreement, still binds the
riparians of East Africa. No riparian, he states, can construct new works or modify the
flow of the Nile without consultation with *and the agreement of* Egypt and the Sudan (p.
133).

5. This stance, which can be summarized "what we need is ours; what we don't need, we
share," underlay the 1957 Canada–United States negotiations on the Waterton and Belly
rivers (see Chapter 1, note 7). It is a predictable assertion of acquired rights. In the case of
Egypt and the Sudan, those rights had been asserted only for two years.

6. The threat was stated in an address to the Sudanese parliament by Minister of Interior

Tayeb Ibrahim Mohamed Khair (Inter Press Third World News Agency, July 5, 1995). For a considered Egyptian analysis see Rushdy Said (1995).

7. In the mid-1970s, the Sudan developed a 300,000 *feddan* cane sugar scheme at Kennana adjacent to the White Nile. As it came "on stream," international oil prices shot up for a second time, raising the costs of pumping and thereby calling into question the economic feasibility of the project.

8. In the summer of 2000 Ethiopia began negotiations with the Sudan for the supply of crude petroleum to Ethiopia.

9. For the deep history of this project, see Collins, 1990; for the best analysis of its likely impact, see Howell et al., 1988.

10. The canal, it turned out, was not raised above the level of the surrounding plains, but below it, thereby obviating the possibility of gravity-flow irrigation.

11. The huge, laser-guided bucket dredger that the French firm CCI brought in to excavate the canal experienced these problems with vertisols (see Collins, 1990: chapter 9).

12. By summer of 2000, Ugandan and Rwandan forces were fighting *each other* in Kisangani in the Congo. The animosity continued after the assassination of Laurent Kabila in early 2001.

CHAPTER 7. UGANDA: EGYPT'S UNWILLING ALLY

1. There is good evidence that the levels attained in 1963–64 had not been achieved for some 20,000 years. The drop over that period had been caused by uplift or tilting in the southwestern part of Lake Victoria. In three years the work of 20,000 was overcome by high rainfall (see Brachi, 1960, and Bishop, 1969; I am grateful to John Sutcliffe for these citations).

2. The smaller Kenyan rivers are the Nzoia, the Yala, the Nyando, the Sonda, and the Gusha/Migori. The Mara originates in Kenya, flows into Tanzania, and then empties into Lake Victoria.

3. Howell notes that the East African démarche implicitly recognized the writ of the 1929 agreement, which, ironically, had already been repudiated by the Sudan.

4. In 1999, with backing from the World Bank and the UNDP, Tecconile was superseded by the Nile Basin Initiative. This transition in fact marked a diminution in Egypt's preponderant role in setting the terms for basin-wide cooperation.

5. So suddenly has the hyacinth spread that only a decade ago a World Bank study could still state, "It is not likely that waterweed will proliferate in Lake Victoria or the Victoria Nile to create problems with power generation at Owen Falls" (World Bank, 1991:76).

6. Prior to 1961–64 the average throughput had been 800 cumsec. It rose to 1,200 cumsec after 1964.

CONCLUSION. LESSONS LEARNED?

1. The proven reserves of petroleum at Bentiu in south-central Sudan are substantial, but it is not clear if they could sustain the economy. Given the volatility in international oil

markets and the vulnerability of the fields and pipeline to attack, oil production and exports would be a frail reed on which to lean.

2. Cited from the organization's web site, *www.nilebasin.org.*

3. If Egypt followed its 1959 criteria in allocating the net benefit of new water management schemes, it could offer upstream riparians to share the costs of projects of common benefit 50–50 but claim only 25 percent of the net water benefit.

Bibliography

Abate, Solomon (1994) *Land Use Dynamics, Soil Degradation and Potential for Sustainable Use in Metu Area, Illubabor Region, Ethiopia.* African Studies Series A 13, Institute of Geography, University of Berne.

Abate, Zewdie (1991) "Planned National Water Policy: A Proposed Case for Ethiopia." International Workshop on Comprehensive Water Resources Management Policies, World Bank, Washington, D.C., June 24–28.

———(1992) *Observed Context of Environment and Development in Ethiopia: An Approach that Needs Development.* London: SOAS, University of London.

———(1994) *Water Resources Development in Ethiopia: An Evaluation of Present Experience and Future Planning Concepts.* Reading: Ithaca Press.

Abdel Ati, Hassan (1992) "The Damming of the River Atbara and Its Downstream Impact." In Michael Darkoh (ed.), *African River Basins and Dryland Crises.* Research Program on Environment and International Security, Reprocentralen, Uppsala, Sweden, pp. 21–44.

Abd el Ghany, H. E. S., and M. Y. H. Elwan (1996) "Environmental Impacts of the Upper Nile Conservation Projects." Paper presented at the Fifth Nile 2002 Conference, Addis Ababa, Feb. 24–28.

Abdel Mageed, Yahya (1994) "The Central Region: Problems and Perspectives." In Peter Rogers and Peter Lydon (eds.), *Water in the Arab World: Perspectives and Prognoses.* Cambridge: Harvard University Press, pp. 101–20.

Abdel Nour, Hassan, and Kidare Mengestu (1994) "Conservation and Development of the Guang, Angereb, and Atbara Catchments: Zeroing in on an Enormous Task." In *Proceedings of the Khartoum Nile 2002 Conference*. Khartoum: African University Printing Press, pp. 651–56.

Abdul Mohsen, Assem (1980) "Egypt, Ethiopia Clash over Nile." *The Middle East* (London) September, p. 70.

Acres International Ltd. (1990) *Proposed Extension to Owen Falls Generating Station: Feasibility Study Report*. UEB and the World Bank.

ADE (For the EU) (1996) *European Strategy to Support Food Security in Ethiopia*. Brussels, February.

Agricultural Policy Committee (Uganda) (1995) *Emergency Action Plan for Control of Water Hyacinth*. Recommendations of the National Technical Committee on Control and Management of Water Hyacinth. Kampala, September.

Ahmed, Siddig (1996) "Release Policies for Multipurpose Reservoirs During Different Climatic Fluctuations." *Water International*, 21, pp. 93–99.

Aiyar, Mani Shankar (1998) "Cheating Tamil Nadu: The Cauvery Authority Is a Fraud on the Farmer of the Delta." *India Today*, Aug. 24, p. 21.

Alemu, Senai (1995a) *The Nile Basin: Data Review and Riparian Issues*. Final Draft Report: Appendices. Washington, D.C.: World Bank, August.

——— (1995b) "Problem Definition and Stakeholder Analysis of the Nile River Basin." Paper presented to the Third Nile 2002 Conference, Arusha, Tanzania, Feb. 13–15.

Allan, J. Anthony (1994) "Policy Issues in Basin-wide Integrated Rivers Development." In *Proceedings of the Khartoum Nile 2002 Conference*. Khartoum: African University Printing Press, pp. 92–108.

——— (1996) "The Political Economy of Water: Reasons for Optimism but Long-term Caution." In J. A. Allan (ed.), *Water, Peace and the Middle East: Negotiating Resources in the Jordan Basin*. London: Tauris, pp. 75–120.

Alvi, Shansul Haque and Nadir Ahmed Elagib (1996) "Study of Hydrology and Drought in the Flood Region of the Sudan." *Water International*, 21, pp. 76–82.

Annexes 2A–2C (1998) "Bank Operational Policies, Procedures, and Good Practices on International Waterways." In Salman M. A. Salman and Laurence Boisson de Chazournes (eds.), *International Watercourses: Enhancing Cooperation and Managing Conflict*. World Bank Technical Paper no. 414. Washington, D.C.: World Bank, pp. 193–202.

Aredo, Dejane (1989) "Famine Causation, Food Aid, and Foreign Financial Assistance to Agriculture in Ethiopia." In Office of the National Committee for Central Planning, "Towards a Food and Nutrition Strategy for Ethiopia." Proceedings of the National Workshop on Food Strategies for Ethiopia, Alemaya University of Agriculture, Dec. 8–12, 1986. Addis Ababa, pp. 58–73.

Asfaw, Gideon (1993) "Water Resources and Regional Development: Implications of Emerging Government Policies." In Zewdie Shibre and Abdulhamid Bedri (eds.), *Regional Development Problems in Ethiopia*, Institute of Development Research (Addis Ababa University), Friedrich Ebert Foundation. Addis Ababa, December, pp. 125–36.

Assafadowlah, M. (1995) "Sharing Transboundary Rivers: The Ganges Tragedy." In Gerald Blake et al. (eds.), *The Peaceful Management of Transboundary Resources*. London and Dordrecht: Graham and Trotman, Martinus Nijhoff, pp. 209–18.

'Auda, 'Abd al-Malik (1993) "Egypt, and Ethiopia, and the Nile." *al-Ahram al-Iqtisadi,* May 30, p. 82 (in Arabic).

Ayalew, Sosina (1994) "Ethiopia's Foreign Policy." *Ethioscope* (Addis Ababa) 1, no. 1, pp. 3–10.

Ayeb, Habib (1998) "Eau et politiques d'aménagement du territoire en Egypte." *Maghreb-Machrek,* no. 162, Oct.–Dec., pp. 69–84.

Badawi, E. F. el-Monshid, Omer el-Awad, and Siddig Ahmed (1997) "Environmental Impact of the Blue Nile Sediment on Reservoirs and Irrigation Canals." Paper presented at the Fifth Nile 2002 Conference, Addis Ababa, Feb. 24–28.

Badr, Marwan (1995) "Le Nil et la nécessaire coopération." *al-Ahram Hebdo,* Feb. 1–7, p. 10.

Bates, Robert (1988) "Contra Contractarianism: Some Reflections on the New Institutionalism." *Politics and Society,* 16, pp. 387–401.

———(1997) *Open-Economy Politics: The Political Economy of the World Coffee Trade.* Princeton: Princeton University Press.

Baumol, William J. (1986) *Superfairness.* Cambridge: MIT Press.

Beaumont, Peter (1994) "The Myth of Water Wars and the Future of Irrigated Agriculture in the Middle East." *International Journal of Water Resources Development,* 10, no. 1, pp. 9–21.

———(2000) "The 1997 UN Convention of the Law of Non-Navigational Uses of International Watercourses: Its Strengths and Weaknesses from a Water Management Perspective and the Need for New Workable Guidelines." *International Journal of Water Resources Development,* 16, no. 4, pp. 475–96.

Becker, Nir, and Naomi Zeitouni (1998) "A Market Solution for the Israeli-Palestinian Water Dispute." *Water International,* 23, no. 4, pp. 238–43.

Bellete, Solomon (1989) "Ethiopia's Agricultural Production Strategies: An Overview." Office of the National Committee for Central Planning, "Towards a Food and Nutrition Strategy for Ethiopia." Proceedings of the National Workshop on Food Strategies for Ethiopia, Alemaya University of Agriculture, Dec. 8–12, 1986. Addis Ababa, pp. 74–96.

Bilen, Ozden (1997) *Turkey and Water Issues in the Middle East.* Ankara: GAP.

Bingham, Gail, Aaron Wolf, and Tim Wohlgenant (1994) *Resolving Water Disputes: Conflict and Cooperation in the United States, the Near East, and Asia.* Arlington, Va.: Irrigation Support Project for Asia and the Near East (ISPAN) and USAID.

Bishop, W. W. (1969) "Pleistocene Stratigraphy in Uganda." *Geological Survey of Uganda,* memoir no. 10, pp. 102–17.

Blake, Gerald H., et al. (eds.) (1995) *The Peaceful Management of Transboundary Resources.* London and Dordrecht: Graham and Trotman, Martinus Nijhoff.

Blue Nile Consultants (Coyne and Bellier, Sir A. Gibb and Partners, Hunting Technical Services, Sir M. MacDonald and Partners) (1978) *Blue Nile Waters Study.* Vols. I–VII. Khartoum, Sudan: Ministry of Irrigation and Hydro-electric Energy.

Boggess, William, Ronald Lacewell, and David Zilberman (1993) "Economics of Water Use in Agriculture." In William Boggess et al. (eds.), *Agricultural and Environmental Resource Economics.* New York: Oxford University Press, pp. 319–91.

Boisson de Chazournes, Laurence (1998) "Elements of a Legal Strategy for Managing International Watercourses: The Aral Sea Basin." In Salman M. A. Salman and Laurence Boisson de Chazournes (eds.), *International Watercourses: Enhancing Cooperation and Managing Conflict.* World Bank Technical Paper no. 414. Washington, D.C.: World Bank, pp. 65–76.

Bourne, Charles B. (1992) "The International Law Commission's Draft Articles on the Law of International Watercourses: Principles and Planned Measures." *Colorado Journal of International Environmental Law,* 3, pp. 65–92.

Brachi, R. M. (1960) "Excavation of a Rock Shelter at Hippo Bay, Entebbe." *Uganda Journal,* 21, no. 1, pp. 62–70.

Brichieri-Colombi, Steven (1996) "Equitable Use and the Sharing of the Nile." Paper presented at the Fourth Annual Nile 2002 Conference, Kampala, Uganda, Feb. 26–29.

Briscoe, John (1999a) "The Changing Face of Water Infrastructure Financing in Developing Countries." *International Journal of Water Resources Development,* 15, no. 3, pp. 301–8.

———(1999b) "The Financing of Hydropower, Irrigation, and Water Supply Infrastructure in Developing Countries." *International Journal of Water Resources Development,* 15, no. 4, pp. 459–92.

Butter, David (1998) "Egypt to Forge Ahead with Nile Water Project." *Middle East Economic Digest,* Jan. 16, pp. 2–3.

Caflisch, Lucius (1998) "Regulation of the Uses of International Watercourses." In Salman M. A. Salman and Laurence Boisson de Chazournes (eds.), *International Watercourses: Enhancing Cooperation and Managing Conflict.* World Bank Technical Paper no. 414. Washington, D.C.: World Bank, pp. 3–16.

Cano, Guillermo (1986) "The Del Plata Basin: Summary Chronicle of Its Development Process and Related Conflicts." In Evan Vlachos (ed.), *The Management of International Rivers.* Laxenburg, Austria: IASA.

Caponera, Dante, and Dominique Alhéritière (1978) "Principles for International Groundwater Law." *Natural Resources Journal,* 18, pp. 589–619.

Chayes, Abram, and Antonia Handler Chayes (1995) *The New Sovereignty: Compliance with International Regulatory Agreements.* Cambridge: Harvard University Press.

Cheesman, Major R. E. (1968) *Lake Tana and the Blue Nile: The Abyssinian Quest* (1st ed., 1938). London: Frank Cass.

Chomchai, Prachoom (1995) "Management of Transboundary Water Resources: The Case of the Mekong." In Gerald Blake et al. (eds.), *The Peaceful Management of Transboundary Resources.* London and Dordrecht: Graham and Trotman, Martinus Nijhoff, pp. 245–60.

Collins, Robert (1990) *The Waters of the Nile: Hydropolitics and the Jonglei Canal, 1900–88.* Oxford, Clarendon Press.

Constable, Michael (1986) *The Degradation of Resources and an Evaluation of Actions to Combat It.* Part II, draft, EHRS, FAO, Addis Ababa.

Constable, Michael, and Deryke Belshaw (1989) "The Ethiopian Highlands Reclamation Study." In Office of the National Committee for Central Planning, "Towards a Food and Nutrition Strategy for Ethiopia." Proceedings of the National Workshop on Food Strategies for Ethiopia, Alemaya University of Agriculture, Dec. 8–12, 1986, Addis Ababa, pp. 142–79.

"Convention on the Law of the Non-Navigational Uses of International Watercourses" (1998) In Salman M. A. Salman and Laurence Boisson de Chazournes (eds.), *Interna-*

tional Watercourses: Enhancing Cooperation and Managing Conflict. World Bank Technical Paper no. 414. Washington, D.C.: World Bank, pp. 173–192.

Cooper, Richard (1997) "A Treaty on Global Climate Change: Problems and Prospects." Weatherhead Center for International Affairs, Harvard University, Working Paper Series no. 97–9, December.

Crow, Ben (1995) *Sharing the Ganges: The Politics and Technology of River Development.* Los Angeles: Sage Publications.

Crow, Ben, and Nirvikar Singh (2000) "Impediments and Innovation in International Rivers: The Waters of South Asia." *World Development,* 28, no. 11, pp. 1907–26.

Damodaran, Ashok (1996) "Cauvery Dispute: Drowned in Politics." *India Today,* January 31, pp. 28–33.

d'Arge, Ralph, and Allen Kneese (1980) "State Liability for International Environmental Degradation: An Economic Perspective." *Natural Resources Journal,* 20, no. 3, pp. 427–50.

Darkoh, Michael (ed.) (1992) *African River Basins and Dryland Crises.* Research Program on Environment and International Security. Reprocentralen, Uppsala, Sweden.

Democratic Republic of the Sudan, Ministry of Irrigation (1978) *Blue Nile Waters Study, Phase IA: Availability and Use of Blue Nile Water.* Vol. 1, *The Main Report.* Coyne and Bellier, Sir Alexander Gibb, HTS, Sir M. MacDonald. Khartoum, April.

Democratic Republic of the Sudan (1979) *Nile Waters Study.* Vol. 1, *Main Report;* vol. 2, *Supporting reports I–III;* vol. 3, *Irrigation;* vol. 4, *Hydroelectric Projects, Hydrology, System Model.*

Deng, Francis (1995) *War of Visions: Conflict of Identities in the Sudan.* Washington, D.C.: Brookings Institution.

Diriba, Getachew (1995) *Economy at the Crossroads: Famine and Food Security in Rural Ethiopia.* Addis Ababa: Care International.

Donovan, Graeham (1996) "Ethiopia: Strategy for Food Security." Draft. Washington, D.C.: World Bank.

Dribidu, Enoch (1996) "Towards Integrated Water Resources Management in Uganda." Paper presented to the Nile 2002 Conference, Addis Ababa, February.

Economic Commission on Africa (ECA) and UNDP (1989) "Report of the Workshop on the Nile Basin Integrated Development Fact Finding Mission." Draft report, NRD/WRNBD/3/89, Addis Ababa.

EELPA (Ethiopian Electric Light and Power Authority) (1982) *Power Planning Study: Main Report* (Acres International, Niagara, Canada), October.

Erlich, Haggai (1994) *Ethiopia and the Middle East.* Boulder. Colo.: Lynne Rienner.

Ethiopia (Government of) (1977) *United Nations Water Conference: Ethiopia.* E/Conf. 70/TP 258, Mar del Plata, Argentina, March 14–25.

Ethiopia, Republic of (1997) *Country Paper-Ethiopia: Water Resources Management of the Nile Basin; Basis for Cooperation.* Paper presented at the Fifth Nile 2002 Conference, Addis Ababa, Feb. 24–28.

Ethiopian Mapping Agency (1988) *National Atlas of Ethiopia.* Addis Ababa.

EVDSA (1992) *A Review of the Water Resources of Ethiopia* (Woodroofe Report). Addis Ababa.

Falkenmark, Malin (1989) *Natural Resource Limits to Population Growth: The Water Perspective.* Gland, Switzerland: International Union for the Conservation of Nature.

FAO (1986) *Ethiopian Highlands Reclamation Study.* Rome: FAO.

———. (1995) *Emergency Control of Water Hyacinth—Lake Victoria: Uganda.* TCP/RAF/2371. Rome, April.

FAO, World Bank, and UNDP (1995) *Water Sector Policy Review and Strategy Formulation: A General Framework.* FAO Land and Water Bulletin. Rome.

Federal Democratic Republic of Ethiopia (1996) "Water Resources Development in Ethiopia." Paper presented at the Fourth Nile 2002 Conference, Kampala, Uganda, Feb. 26–29.

Field, Michael (1973) "Developing the Nile." *World Crops,* 25, no. 1, pp. 11–15.

Fisher, Franklin (1995) "The Economics of Water Dispute Resolution, Project Evaluation, and Management: An Application to the Middle East." *Water Resources Development,* 11, no. 4, pp. 377–90.

Fisseha, Yimr (1981) "State Succession and the Legal Status of International Rivers." In R. Zacklin and L. Calfisch (eds.), *The Legal Regime of International Rivers and Lakes.* The Hague: Martinus Nijhoff, pp. 177–202.

Flint, Courtney G. (1995) "Recent Developments of the International Law Commission Regarding International Watercourses and Their Implications for the Nile River." *Water International,* 20, pp. 197–204.

Garang de Mabior, John (1981) *Indentifying, Selecting, and Implementing Rural Development Strategies for Socio-Economic Development in the Jonglei Project Area, Southern Region.* Ph.D. diss., Iowa State University.

Garstin, Sir. William (1904) *Report upon the Basin of the Upper Nile with Proposals for the Improvement of that River.* Cairo: National Printing Department.

Gelb, Alan, and associates (1988) *Oil Windfalls: Blessing or Curse?* New York: Oxford University Press.

Gilpin, Robert (1981) *War and Change in World Politics.* Cambridge: Cambridge University Press.

Glieck, Peter (1996) "Basic Water Requirements for Human Activities: Meeting Basic Needs." *Water International,* 21, pp. 83–92.

Godana, Bonaya (1985) *Africa's Shared Water Resources: Legal and Institutional Aspects of the Nile, Niger, and Senegal River Systems.* Boulder, Colo.: Lynne Rienner.

Goldberg, David (1995) "World Bank Policy on Projects on International Waterways in the Context of Emerging International Law and the Work of the International Law Commission." In Gerald Blake, et al. (eds.), *The Peaceful Management of Transboundary Resources.* London and Dordrecht: Graham and Trotman, Martinus Nijhoff, pp. 153–66.

Goldsmith, Edward, and Nicholas Hildyard (1984) *The Social and Environmental Effects of Large Dams, Volume 1: Overview.* Bordeaux, France, and Powys, Wales: European Ecological Action Group.

Gould, G. (1988) "Water Rights Transfers and Third-Party Effects." *Land and Water Law Review,* 23, pp. 1–41.

Guariso, Giorgio, and Dale Whittington (1987) "Implications of Ethiopian Water Development for Egypt and the Sudan." *Water Resources Development,* 3, no. 2, pp. 105–14.

Gupta, Rajiv K. (2001) "River Basin Management: A Case Study of Narmada Valley Development with Special Reference to the Sardar Sarovar Project in Gujarat, India." *International Journal of Water Resources Development,* 17, no. 1, pp. 55–78.

Haas, Peter (1990) *Saving the Mediterranean: The Politics of International Environmental Co-operation.* New York: Columbia University Press.

———(1993) "Epistemic Communities and the Dynamics of International Co-operation." In Volker Rittberger (ed.), *Regime Theory and International Relations.* Oxford: Clarendon Press, pp. 168–201.

Haggai, Erlich (1994) *Ethiopia and the Middle East.* Boulder, Colo.: Lynne Rienner.

Haile-Mariam, Assefa (1995) "Population Growth and Environmental Situation in Ethiopia." *Ethiopian Herald* (Addis Ababa), April 28, p. 2.

Hailie Selassie I (1976) *My Life and Ethiopia's Progress, 1892–1937.* Translated and annotated by Edward Ullendorf. Oxford: Oxford University Press.

———(1994) *My Life and Ethiopia's Progress,* vol. II. Edited by Harold Marcus. East Lansing: Michigan State University Press.

Hamidi, Ibrahim (1998) "The Tigris-Euphrates Waters File Elicits Consensus in the League." *al-Hayat,* August 16, p. 3 (in Arabic).

Hardin, Russell (1982) *Collective Action.* Baltimore: Published for Resources for the Future, Inc., by Johns Hopkins University Press.

Haynes, Kingsley, and Dale Whittington (1981) "International Management of the Nile—Stage Three?" *Geographical Review,* 71, no. 1, January, pp. 17–32.

Herbst, Jeffrey (2000) *States and Power in Africa.* Princeton: Princeton University Press.

Hilmy, N. A. (1978) "Some Legal Questions About Irrigation from the River Nile." *Revue Egyptienne de Droit International,* 34, pp. 123–48.

Hirji, Rafik, and David Grey (1997) "Managing International Waters n Africa: Process and Progress." Paper presented at the Fifth Nile 2002 Conference, Addis Ababa, Feb. 24–28; reprinted in Salman M. A. Salman and Laurence Boisson de Chazournes (eds.), *International Watercourses: Enhancing Cooperation and Managing Conflict.* World Bank Technical Paper no. 414. Washington, D.C.: World Bank, pp. 77–100.

Hoben, Allan (1995) "Paradigms and Politics: The Cultural Construction of Environmental Policy in Ethiopia." *World Development,* 23, no. 6, pp. 1007–22.

Homer-Dixon, Thomas (1994) "Environmental Scarcities and Violent Conflict: Evidence from Cases." *International Security,* 19, no. 3, pp. 5–40.

Hopper, David (1976) "The Development of Agriculture in Developing Countries." *Scientific American,* 235, no. 3, pp. 196–205.

Hori, Hiroshi (1993) "Development of the Mekong River Basin: Its Problems and Future Prospects." *Water International,* 18, no. 2.

Howard, Julie, et al. (1995) "Toward Increased Domestic Cereals Production in Ethiopia." Food Security Project, Ministry of Economic Development and Cooperation, Working Paper 3. Addis Ababa.

Howell, Paul (1994) "East Africa's Water Requirements: The Equatorial Nile Project and the Nile Waters Agreement of 1929." In Paul Howell and J. A. Allan (eds.), *The Nile: Sharing a Scarce Resource.* Cambridge: Cambridge University Press, pp. 81–107.

Howell, Paul, and J. A. Allan (eds.) (1994) *The Nile: Sharing a Scarce Resource.* Cambridge: Cambridge University Press.

Howell, Paul, and Michael Lock (1994) "The Control of the Swamps of the Southern Sudan: Drainage Schemes, Local Effects, and Environmental Constraints on Remedial Development in the Flood Region." In Paul Howell and J. A. Allan (eds.), *The Nile: Sharing a Scarce Resource.* Cambridge: Cambridge University Press, pp. 243–80.

Howell, Paul, Michael Lock, and Stephen Cobb (eds.) (1988) *The Jonglei Canal: Impact and Opportunity.* New York: Cambridge University Press.

Huffaker, Ray, Norman Whittlesey, and Joel Hamilton (2000) "The Role of Prior Appropriation in Allocating Water Resources in the Twenty-first Century." *International Journal of Water Resources Development,* 16, no. 2, pp. 265–73.

Hurst, H. E. (1950) *The Nile Basin.* Vol. VIII, *The Hydrology of the Sobat and White Nile and the Topography of the Blue Nile and Atbara.* Cairo: Ministry of Public Works, Government Press.

Hurst, H. E., R. P. Black, and Y. M. Simaika (1951) *The Nile Basin.* Vol. VII, *The Future Conservation of the Nile.* Cairo: National Printing Office.

——— (1959) *The Nile Basin.* Vol. IX, *The Hydrology of the Blue Nile and Atbara and of the Main Nile to Aswan, with Some Reference Projects.* Cairo: Ministry of Public Works, General Organization for Government Printing Offices.

——— (1966) *The Nile Basin.* Vol. X, *The Major Nile Projects.* Cairo: Ministry of Irrigation, General Organization for Government Printing Offices.

Hurst, H. E., and P. Phillips (1931) *The Nile Basin.* Vol. 1, *General Description of the Basin, Meteorology, Topography of the White Nile Basin.* Cairo: Ministry of Public Works, Government Press.

Hutchinson, Sharon (1996) *Nuer Dilemmas: Coping with Money, War, and the State.* Berkeley: University of California Press.

Hydromet (1974) *Hydrometeorological Survey of the Catchments of Lake Victoria, Kyoga, and Albert.* UNDP/World Meteorological Organization, RAF 66–025, Geneva.

——— (1981) *Hydrometeorological Survey of the Catchments of Lake Victoria, Kyoga, and Mobutu Sese Seko: Project Findings and Recommendations.* RAF/73/001, Geneva.

IGADD (1991) "Description of ndvi Processing Methodology Used by the Project: An Application to Ethiopia During the Last Three Years." Working Paper No. 1 (GCPS/RAF/256/ITA), Nairobi, March 4–5.

Isaac, Jad, and Hillel Shuval (eds.) (1994) *Water and Peace in the Middle East.* Amsterdam: Elsevier.

Ja'ali, al-Bukhari 'Abdullah al- (1980) *The Border Dispute Between Sudan and Ethiopia.* Kuwait: Matba' al-Khalij.

Jansonius, Jan (1989) "Food Security Systems." In Office of the National Committee for Central Planning, "Towards a Food and Nutrition Strategy for Ethiopia." Proceedings of the National Workshop on Food Strategies for Ethiopia, Alemaya University of Agriculture, Dec. 8–12, 1986; Addis Ababa, pp. 97–127.

Jovanovic, D. (1985) "Ethiopian Interests in the Division of the Nile Waters." *Water International,* 10, pp. 82–85.

Kabanda, B. K., and P. O. Kahangire (1991) "Monitoring, Forecasting and Simulation of River Basins for Water Resources Development in Uganda: A Review of Practice and Constraints." FAO Seminar on Monitoring, Forecasting, and Simulation of River Basins for Agricultural Production, Bologna, March 17–23.

Kagera Basin Organization (Secretariat) (1979) *General Background Information on the Planning of the Kagera River Basin.* Kigali, April.

Kahangire, Patrick, and Enoch Dribidu (1995) "Rapid Assessment of Uganda's Water Resources and Demands." Paper presented to the Third Nile 2002 Conference, Arusha, Tanzania, Feb. 13–17.

Kally, Elisha (1994) "Costs of Inter-regional Conveyance of Water and Costs of Sea Water Desalinization." In Jad Isaac and Hillel Shuval (eds.), *Water and Peace in the Middle East.* Amsterdam: Elsevier, pp. 289–300.

Karshenas, Massoud (1994) "Environment, Technology, and Employment: Towards a New Definition of Sustainable Development." *Development and Change,* 25, pp. 723–56.

Kaufman, Les (1995) "Brief Notes on the Water Hyacinth Crisis and What to Do About It." (Consultant to the World Bank.) Boston University, unpublished.

Kennett, Steven A. (1991) *Managing Interjurisdictional Waters in Canada: A Constitutional Analysis.* Calgary: Faculty of Law, University of Calgary.

Killion, Tom (1992) "Refugee and Environmental Change on the Sudano-Ethiopian Frontier: A History of Wad el-Hileau Refugee Camp, 1967–87." Sixth Conference of Northeastern African Studies, East Lansing, Michigan, April.

Kite, G. W. (1981) "Recent Changes in Level of Lake Victoria." *Hydrological Sciences,* 26, no. 3, pp. 233–43.

Kliot, Nurit (1994) *Water Resources and Conflict in the Middle East.* London: Routledge.

Kloos, Helmut (1991) "Peasant Irrigation Development and Food Production in Ethiopia." *Geographical Journal,* 157, November, pp. 295–306.

Kolars, John, and William Mitchell (1991) *The Euphrates River and the Southeast Anatolia Development Project.* Carbondale: Southern Illinois University Press.

Koren, Victor, Curtis Barnett, and Wulf Khohn (1994) "Satellite Based Forecasting and Monitoring System for the Nile Basin." In *Proceedings of the Khartoum Nile 2002 Conference.* Khartoum: African University Printing Press, pp. 75–80.

Krasner, Stephen (1983a) "Regimes and the Limits of Realism." In Stephen Krasner (ed.), *International Regimes.* Ithaca, N.Y.: Cornell University Press, 355–68.

———(1983b) "Structural Causes and Regime Consequences: Regimes as Intervening Variables." In Stephen Krasner (ed.), *International Regimes.* Ithaca, N.Y.: Cornell University Press, 1–22.

Krishna, Raj (1998) "The Evolution and Context of the Bank Policy for Projects on International Waterways." In Salman M. A. Salman and Laurence Boisson de Chazournes (eds.), *International Watercourses: Enhancing Cooperation and Managing Conflict.* World Bank Technical Paper no. 414. Washington, D.C.: World Bank, pp. 31–44.

Lako, George Tombe (1992) "The Jonglei Canal Scheme as a Socio-Economic Factor in the Civil War in the Sudan." In Michael Darkoh (ed.), *African River Basins and Dryland Crises.* Research Program on Environment and International Security. Reprocentralen, Uppsala, Sweden, pp. 45–58.

Lasswell, Harold J. (1936) *Politics: Who Gets What, When, and How.* New York: P. Smith.

Lee, Donna J., and Ariel Dinar (1996) "Integrated Models of River Basin Planning, Development, and Management." *Water International,* 21, pp. 213–22.

Lee, Gwang-Man, Muleggeta Seid, and Yimam Yimer (1996) "Small-Scale Area-Based Water Harvesting Approach for Agricultural Activities in Highlands of Ethiopia." Paper presented to the Fifth Nile 2002 Conference, Addis Ababa, Feb. 24–28.

Legro, Jeffrey, and Andrew Moravcsik (1998) "Is Anybody Still a Realist?" Weatherhead Center for International Affairs, Harvard University, Working Paper no. 98-14, October.

Le Marquand, David (1977) *International Rivers: The Politics of Cooperation.* Vancouver: University of British Columbia Press.

Lichbach, Mark I. (1996) *The Cooperator's Dilemma.* Ann Arbor: University of Michigan Press.

Lilienthal, David (1953) *TVA: Democracy on the March.* Chicago: Quadrangle Books (1966 paperback edition).

Lowi, Miriam (1993) *Water and Power: The Politics of a Scarce Resource in the Jordan River Basin.* New York: Cambridge University Press.

Mann, Oscar (1977) *The Jonglei Canal: Environmental and Social Aspects.* Nairobi: Environment Liaison Center.

Marcus, Amy (1997) "Water Fight: Egypt Faces Problem It Has Long Dreaded: Less Control of the Nile." *Wall Street Journal,* Aug. 22.

Marty, Frank (1997) *International River Management: The Political Determinants of Success and Failure.* Studien zur Politikwissenschaft, no. 305, Institut für Politikwissenschaft, Zurich.

Mbugua, Joseph (1987) "Water Resource Use Efficiency in Lake Basin Area of Kenya." In George Ruigu and Mandivamba Rukuni (eds.), *Irrigation Policy in Kenya and Zimbabwe.* Institute for Development Studies, University of Nairobi, pp. 42–48.

McCaffrey, Stephen (1989) "The Law of International Watercourses: Some Recent Developments and Unanswered Questions." *Denver Journal of International Law and Policy,* 17, pp. 505–26.

———(1998) "The UN Convention on the Law of the Non-Navigational Uses of International Watercourses: Prospects and Pitfalls." In Salman M. A. Salman and Laurence Boisson de Chazournes (eds.), *International Watercourses: Enhancing Cooperation and Managing Conflict.* World Bank Technical Paper no. 414. Washington, D.C.: World Bank, pp. 17–28.

McCann, James (1981) "Britain, Ethiopia, and the Lake Tana Project; 1922–35." *International Journal of African Historical Studies,* 14, pp. 667–99.

———(1990) "A Dura Revolution and Frontier Agriculture in Northwest Ethiopia, 1898–1920." *Journal of African History,* 31, pp. 121–34.

McCully, Patrick (1998) *Silenced Rivers.* London: Orient Longman.

McDougall, I. A. (1971) "The Development of International Law with Respect to Trans-Boundary Water Resources: Co-operation for Mutual Advantage or Continentalism's Thin Edge of the Wedge?" *Osgoode Hall Law Journal,* 9, no. 2, pp. 261–311.

Micklin, Philip (1991) "The Water Management Crisis in Soviet Central Asia." *The Carl Beck Papers in Russian and East European Studies,* no. 905, University of Pittsburgh.

Ministry of Agriculture (Government of Socialist Ethiopia) (1984) *General Agricultural Survey.* Addis Ababa.

———(1987) *Assessment of Agricultural Land Suitability in South Eastern, Southern, South Western, and Western Ethiopia.* Vol. 1, *Main Report.* Australian Agricultural Consulting and Management Company. Addis Ababa, June.

Ministry of Foreign Affairs (Government of Socialist Ethiopia) (1993) "Treaties and Agreements on the Nile Waters: 1894–1993." Addis Ababa.

Ministry of Information (Kingdom of Ethiopia) (1969) *Patterns of Progress: Power and Irrigation in Ethiopia.* Addis Ababa.

Ministry of Mines and Energy (Government of Socialist Ethiopia) (1986) *Hydroenergy Resources: Technical Report no. 2.* CESEN-ANSALDO/ Finnmecanica Group, Addis Ababa.

Ministry of Natural Resource Development and Environmental Protection (Transitional Government of Ethiopia) (1994) "Framework for Cooperation Between the Nile River Co-Basin States." Paper presented at the Second Nile 2002 Conference, Khartoum, Jan. 29-31.

Ministry of Public Works (1920) *Nile Control,* vol. 1 (Sir Murdoch MacDonald report). Cairo: Government Press.

Ministry of Water Resources (1995) *Baro-Akobo River Basin Integrated Development Master Plan Project* (Draft Inception Report: Volume I, Main Report). New York and Warwick U.K.: TAMS-ULG Consultants.

———(1996) *Tekeze Medium Hydropower Project* (Pre-feasibility Study: Executive Summary). Howard Humphreys, Coyne and Bellier, Rust, Kennedy and Dorkin. Draft. Addis Ababa, August.

Mitchell, Bruce (1990) *Integrated Water Management: International Experience and Perspectives.* London: Belhaven Press.

Moghraby, Asim el- (1997) "Water Management in the Sudan." Paper presented at the Seventeenth Annual Meeting of the International Association for Impact Assessment, New Orleans, May.

Nanda, Ved P. (1995) *International Environmental Law and Policy.* Irvington-on-Hudson, N.Y.: Transaction.

Natural Resources Group (1995) *Tekeze River Basin Integrated Master Plan Project,* Agricultural Status Report. Netherlands Engineering Consultants. N.p., April.

Nile Basin Initiative (1999) *Policy Guidelines for the Nile Basin Strategic Action Program.* Entebbe.

North, Douglass (1990) *Institutions, Institutional Change, and Economic Performance.* New York: Cambridge University Press.

Ntale, Henry (1996) "Lake Kyoga, the Nile 'Green' Lake That Is Dying Unnoticed." Paper presented to the Fourth Nile 2002 Conference, Kampala, Feb. 26–29.

Ntambirewki, John, and Akiiki Mujaju (1993) "International, Legal, and Institutional Aspects of the Uganda Water Action Plan." Water Development Department. Kampala, June.

Ochieng, Philip (1996) "Time to Revise Pact Controlling Waters of the Nile." *East African* (Nairobi), March 11–17, p. 10.

Office of the National Committee for Central Planning (1989) "Towards a Food and Nutrition Strategy for Ethiopia." *Proceedings of the National Workshop on Food Strategies for Ethiopia.* Alemaya University of Agriculture, Dec. 8–12, 1986; Addis Ababa.

O'Hara, Sarah L. (2000) "Central Asia's Water Resources: Contemporary and Future Management Issues." *International Journal of Water Resources Deveopment,* 16, no. 3, pp. 423–42.

Ohlsson, Leif (ed.) (1998) *Hydropolitics: Conflict over Water as a Development Constraint.* London: Zed Books.

Okidi Odidi, Charles (1979) "International Laws and the Lake Victoria and Nile Basins." In Charles Okidi Odidi (ed.), *Natural Resources Development of the Lake Victoria Basin of Kenya.* Nairobi: Institute for Development Studies.

———(1986) *Development and the Environment in the Kagera River Basin Under the Rusumo Treaty.* Discussion Paper no. 284, Institute for Development Studies, University of Nairobi, September.

———(1987) "Irrigation Activities and Institutions in Kenya's Lake Victoria Basin." In George Ruigu and Mandivamba Rukuni (eds.), *Irrigation Policy in Kenya and Zimbabwe.* Nairobi: Institute for Development Studies, pp. 267–307.

Olson, Mancur (1971) *The Logic of Collective Action: Public Goods and the Theory of Groups.* Harvard Economic Studies, 124. Cambridge: Harvard University Press.

Osborne, Theresa (1996) "Seasonality and Market News in African Staple Food Markets: Application of Commodity Price Theory to Ethiopian Grain." Research Program in Development Studies, Princeton University, Nov. 15.

Ostrom, Elinor (1991) *Governing the Commons: The Evolution of Institutions for Collective Action.* New York: Cambridge University Press.

———(1992) *Crafting Institutions for Self-Governing Irrigation Systems.* San Francisco: ICS Press.

Overseas Development Administration (ODA) (1993) *A Review and Update of the Water Balance of Lake Victoria in East Africa.* London: ODA Report 93/3, March.

Pankhurst, Richard (1990) *A Social History of Ethiopia.* Institute of Ethiopian Studies, Addis Ababa University.

———(1992) "The History of Deforestation and Afforestation in Ethiopia Prior to World War II." *EJDR,* 2, no. 2, pp. 59–77.

Piper, B. S., D. T. Plinston, and J. V. Sutcliffe (1986) "The Water Balance of Lake Victoria." *Hydrological Sciences,* 31, no. 1, pp. 25–37.

"Power to the People" (1998) *The Economist,* March 28, pp. 61–63.

Proceedings of the Khartoum Nile 2002 Conference (1994). Khartoum: African University Printing Press.

Putnam, Robert (1988) "Diplomacy and Domestic Politics: The Logic of Two-level Games." *International Organization,* 42, pp. 427–60.

Raffer, Kunibert (1997) "Helping Southern Net Food Importers After the Uruguay Round: A Proposal." *World Development,* 25, no. 11, pp. 1901–7.

Rangely, Robert, et al. (1994) *International River Basin Organizations in Sub-Saharan Africa.* World Bank Technical Paper no. 250, Africa Technical Department Series. Washington, D.C.

Rawls, John (1971) *A Theory of Justice.* Cambridge: Belknap Press of Harvard University Press.

Republic of the Sudan (1997) "Country Paper: The Republic of the Sudan." Paper presented to the Fifth Nile 2002 Conference, Addis Ababa, Feb. 24–28.

Rogers, Peter (1991) "International River Basins: Pervasive Unidirectional Externalities." Paper presented at the University of Siena, Conference on the Economics of Transnational Commons, April 25–27.

Rose, Gideon (1998) "Neoclassical Realism and Theories of Foreign Policy." *World Politics,* 51, no. 1, pp. 144–72.

Rosegrant, Mark (1995) "Water Transfers in California: Potentials and Constraints." *Water International,* 20, no. 2, pp. 72–87.

Rosegrant, Mark, and Hans Binswanger (1994) "Markets in Tradable Water Rights: Potential for Efficiency Gains in Developing Country Water Resource Allocation." *World Development,* 22, no. 11, pp. 1613–25.

Ruigu, George, and Mandivamba Rukuni (eds.) (1987) *Irrigation Policy in Kenya and Zimbabwe.* Institute for Development Studies, University of Nairobi.

Runge, C. Ford, and Benjamin Senauer (2000) "A Removable Feast." *Foreign Affairs,* 79, no. 3, pp. 39–51.

Said, Rushdy (1995) "Can Sudan Obstruct the Waters of the Nile?" *al-Musawwar,* July 7, pp. 18–1976.

Salman, Salman M. A. (1998) "Sharing the Ganges Waters Between India and Bangladesh: An Analysis of the 1996 Treaty." In Salman M. A. Salman and Laurence Boisson de Chazournes (eds.), *International Watercourses: Enhancing Cooperation and Managing Conflict.* World Bank Technical Paper no. 414. Washington, D.C.: World Bank, pp. 127–154.

Salman, M. A. Salman, and Laurence Boisson de Chazournes (eds.) (1998) *International Watercourses: Enhancing Cooperation and Managing Conflict.* World Bank Technical Paper no. 414. Washington, D.C.

Sands, Philippe (1998) "Watercourses, Environment and the International Court of Justice: The Gabcikovo-Nagymaros Case." In Salman M. A. Salman and Laurence Boisson de Chazournes (eds.), *International Watercourses: Enhancing Cooperation and Managing Conflict.* World Bank Technical Paper no. 414. Washington, D.C.: World Bank, pp. 103–26.

Schelling, Thomas C. (1997) "The Cost of Combating Global Warming." *Foreign Affairs,* 76, no. 6, pp. 8–14.

Scott, James C. (1998) *Seeing Like a State: How Certain Schemes to Improve the Human Condition Have Failed.* New Haven: Yale University Press.

Sen, Amartya (1981) *Poverty and Famines: An Essay on Entitlements and Deprivation.* New York: Oxford University Press.

Seyoum, Senait, Edgar Richardson, Patrick Webb, Frank Riley, and Yisehac Yohannes (1995) "Analysing and Mapping Food Insecurity: An Exploratory 'cart' Methodology Applied to Ethiopia." IFPRI (Final report to USAID) PIO/T 698-0483-3-3613025, May 1.

Shady, Aly, Ahmad M. Adam, and Kamal Ali Mohamed (1994) "The Nile 2002: The Vision Toward Cooperation in the Nile Basin." *Water International,* 19, pp. 77–81.

Shatanawi, Muhammad, and Odeh Al-Jayousi (1995) "Evaluating Market-oriented Water Policies in Jordan: A Comparative Study." *Water International,* 20, no. 2, pp. 88–96.

Shibre, Zewdie, and Abdulhamid Bedri (eds.) (1993) *Regional Development Problems in Ethiopia.* Institute of Development Research (Addis Ababa University), Friedrich Ebert Foundation. Addis Ababa, December.

Smith, Scot E., and Hussam al-Rawahy (1990) "The Blue Nile: Potential for Conflict and Alternatives for Meeting Future Demands." *Water International,* 15, pp. 217–22.

Supalla, Raymond (2000) "A Game Theoretic Analysis of Institutional Arrangements for Platte River Management." *International Journal of Water Resources Development,* 16, no. 2, pp. 253–64.

Sutcliffe, J. V., and J. B. C. Lazenby (1990) "Hydrological Data Requirements for Planning and Management." Paper presented at the conference on The Nile, Royal Geographical Society, and SOAS, University of London, May 2–3.

Székely, Alberto (1990) "'General Principles' and 'Planned Measures' Provisions in the International Law Commission's Draft Articles on the Non-Navigational Uses of International Water Courses: A Mexican Point of View." *Colorado Journal of International Environmental Law and Policy,* 3, pp. 93–101.

Taylor, Michael (1990) "Cooperation and Rationality: Notes on the Collective Action Problem and Its Solutions." In Karen S. Cook and Margaret Levi (eds.), *The Limits of Rationality.* Chicago: University of Chicago Press, pp. 222–39.

Thesiger, Wilfred (1987) *The Life of My Choice.* Glasgow: HarperCollins.

Tilahun, Wondimneh (1979) *Egypt's Imperial Aspirations over Lake Tana and the Blue Nile.* Addis Ababa, Addis Ababa University.

Transitional Government of Ethiopia (1993) *Directives for Disaster Prevention and Management.* (National Disaster Prevention and Preparedness Committee). Addis Ababa, October.

Tvedt, Terje (1991) "The Management of Water and Irrigation: The Blue Nile." Paper presented at the Workshop on the Prospects for Peace, Recovery, and Development in the Horn of Africa, Institute of Social Studies, The Hague, Feb. 19–23.

———(1992) "Non-implemented Plans as a Barrier to Development: The Case of the Jonglei Project in the Southern Sudan." In Michael Darkoh (ed.), *African River Basins and Dryland Crises.* Research Program on Environment and International Security, Reprocentralen. Uppsala, Sweden, pp. 59–78.

Twongo, T., and J. S. Balirwa (1995) "The Water Hyacinth Problem and the Biological Control Option in the Highland Region of the Upper Nile Basin—Uganda's Experience." Paper presented to the Third Nile 2002 Conference, Arusha, Tanzania, Feb. 13–17.

Uganda: Ministry of Natural Resources (1995) *Uganda Water Action Plan: Main Report.* Directorate of Water Development, doc. 005, Kampala.

Uganda: Water Development Department/DANIDA (1993) *Uganda Water Action Plan: Draft Technical Report #3.* Kampala, June.

Uganda Gazette (1995) "The National Environment Statute, 1995." Supplement, 38, no. 21, May.

United Nations (1963) "Agreement Between the Republic of the Sudan and the United Arab Republic for the Full Utilization of the Nile Waters." *UN Treaty Series,* no. 6519. New York, pp. 64–76.

———(1978) *Register of International Rivers.* Department of Economic Social Affairs. Oxford: Pergamon Press.

UNDP (1989) *Nile Basin Integrated Development: Fact Finding Mission Report.* RAF/86/003-RAB/86/014, July.

UNDP and FAO (1988) *Master Land Use Plan, Ethiopia: Main Report.* AG/ETH/82/010, Technical Report 1. Rome.

UN General Assembly (1997) *Convention on the Law of the Non-navigational Uses of International Watercourses.* Resolution A/RES/51/229. New York, July 8.

UNICEF (1995) *Operation Lifeline Sudan: Review of 1995 Activities.* Nairobi, UNICEF.

U.S. Department of the Interior/Bureau of Reclamation (1964) *Land and Water Resources of the Blue Nile Basin: Ethiopia.* Volume 2: *Plans and Estimates.* Appendix III—Hydrology. Washington, D.C.

———(1984) *Nile Waters Study Evaluation.* Prepared for the Democratic Republic of the Sudan. Washington, D.C., November.

Utton, Albert (1996) "Which Rule Should Prevail in International Water Disputes: That of Reasonableness or That of No Harm?" *Natural Resources Journal,* 36, no. 3, pp. 635–41.

Venema, Henry David, and Eric J. Schiller (1995) "Water Resources Planning for the Senegal River Basin." *Water International,* 20, no. 2, pp. 61–70.

Vinogradov, Sergei (1992) "Observations on the International Law Commission's Draft Rules on the Non-Navigational Uses of International Water Courses: 'Management and Domestic Remedies.'" *Colorado Journal of International Environmental Law and Policy,* 3, pp. 235–59.

Walker, D.C. (1974) "Investment in an Ethiopian Valley." *Geographical Magazine,* 46, pp. 714–18.

WAPCOS (Water and Power Consultancy Services) (1990) *Preliminary Water Resources Development Plan for Ethiopia.* India.

Waterbury, John (1976) "The Sudan in Quest of a Surplus." *American Universities Field Staff Reports,* 21 (Africa Series), nos. 8–10.

———(1977) "Yes, California, There Is a Water Crisis: Observations on the UN Water Conference." *American Universities Field Staff Reports,* 21 (South America Series), no. 1.

———(1979) *Hydropolitics of the Nile Valley.* Syracuse, N.Y.: Syracuse University Press.

———(1982) "Riverains and Lacustrines: Toward International Cooperation in the Nile Basin." Discussion Paper no. 107, Research Program in Development Studies, Princeton University.

———(1994) "Transboundary Water and the Challenge of International Cooperation in the Middle East." In Peter Rogers and Peter Lydon (eds.), *Water in the Arab World.* Cambridge: Division of Applied Sciences, Harvard University, pp. 39–64.

———(1996) "Socio-Economic Development Models for the Nile Basin." Paper presented to the Fourth Nile 2002 Conference, Kampala, Uganda, Feb. 26–29.

———(1997) "Is the Status Quo in the Nile Basin Viable?" *Brown Journal of World Affairs,* 4, no. 1, pp. 287–98.

———(1997) "Between Unilateralism and Comprehensive Accords: Modest Steps Toward Cooperation in International River Basins." *Water Resources Development,* 13, no. 3, pp. 277–90.

Waterbury, John, and Dale Whittington (1998) "Playing Chicken on the Nile: The Implications of Microdam Development in the Ethiopian Highlands and Egypt's New Valley Project." *Natural Resources Forum,* 22, no. 3, pp. 155–64.

Water Development Department (Uganda) (1993) *Uganda Water Action Plan: Draft Technical Report #3.* DANIDA, Kampala, June.

Webb, Patrick, Joachim von Braun, and Yisehac Yohannes (1992) "Famine in Ethiopia: Policy Implications of Coping Failure at National and Household Levels." International Food Policy Research Institute, Washington, D.C.

Whittington, Dale, and Giorgio Guariso (1983) *Water Management Models in Practice: A Case Study of the Aswan High Dam.* Amsterdam: Elsevier Scientific.

Whittington, Dale, and Kingsley Haynes (1985) "Nile Water for Whom? Emerging Conflicts in Water Allocation for Agricultural Expansion in Egypt and the Sudan." In Peter Beaumont and Keith McLaughlin (eds.), *Agricultural Development in the Middle East.* London: John Wiley & Sons, pp. 125–50.

Whittington, Dale, John Waterbury, and Elizabeth McClelland (1994) "Towards a New Nile Waters Agreement." *Ethioscope* (Addis Ababa), 1, no. 1, pp. 11–17.

Wilson, Gail (1967) *Owen Falls: Electricity in a Developing Country.* Nairobi: East African Publishing House.

Wittfogel, Karl (1957) *Oriental Despotism: A Comparative Study of Total Power.* New Haven: Yale University Press.

Wolde-Mariam, Dender (1989) "Food Policy Objectives in the Ten-Year Perspective Plan." In Office of the National Committee for Central Planning, "Towards a Food and Nutrition Strategy for Ethiopia." *Proceedings of the National Workshop on Food Strategies for Ethiopia.* Alemaya University of Agriculture, Dec. 8–12, 1986; Addis Ababa, pp. 9–25.

Wolde-Mariam, Mesfin (1972) *An Introductory Geography of Ethiopia.* Addis Ababa: Berhansena Selam H.S.I. Printing Press.

Wolf, Aaron T. (1996) "Middle East Water Conflicts and Directions for Conflict Resolution." International Food Policy Research Institute Discussion Paper 12. Washington, D.C., March.

———(1997) "From Rights to Needs: 'Equitable' Allocations in International Water Agreements." Paper presented at the Fifth Nile 2002 Conference, Addis Ababa, Feb. 24–28.

Wolf, Aaron T., and Ariel Dinar (1994) "Middle East Hydropolitics and Equity Measures for Water-Sharing Agreements." *Journal of Social, Political, and Economic Studies,* 19, no. 1, pp. 69–93.

Wolf, Aaron T., et al. (1999) "International Rivers of the World." *International Journal of Water Resources Development,* 15, no. 4, pp. 387–428.

World Bank (1969) *Ethiopia: Appraisal of the Finchaa Hydroelectric Project of the Ethiopian Electric Light and Power Authority.* Report no. PU-9a. Washington, D.C., April 10.

———(1976) *Ethiopia: Finchaa Hydroelectric Project: Project Performance Audit Report.* Report no. 1102, Operations Evaluation Department. Washington, D.C., March 23.

———(1979) *Kagera River Basin: Regional Development.* Report no. 2812, East Africa Regional Office. Washington, D.C., December.

———(1990) *Operational Directive: Projects on International Waterways,* OD 750. Washington, D.C., April.

———(1991) *Staff Appraisal Report: Uganda; Third Power Project.* Report no. 9153-UG, Industry and Energy Division, Eastern Africa Department. Washington, D.C., May 29.

Young, Oran (1994) *International Governance: Protecting the Environment in a Stateless Society.* Ithaca, N.Y.: Cornell University Press.

Zewde, Bahru (1976) "Relations Between Ethiopia and the Sudan on the Western Ethiopian Frontier, 1898–1935." D.Phil. thesis, University of London.

———(1994) *A History of Modern Ethiopia: 1855–1974.* London: James Currey.

Index